Essays in Biochemistry

Other recent titles in the Essays in Biochemistry series

Essays in Biochemistry volume 42: The Biochemical Basis of the Health Effects of Exercise
edited by A. Wagenmakers
2006 ISBN-10 1 85578 159 X
ISBN-13 978 1 85578 159 7

Essays in Biochemistry volume 41: The Ubiquitin–Proteasome System
edited by R.J. Mayer and R. Layfield
2005 ISBN-10 1 85578 153 0
ISBN-13 978 1 85578 153 5

Essays in Biochemistry volume 40: The Nuclear Receptor Superfamily
edited by I.J. McEwan
2004 ISBN-10 1 85578 150 6
ISBN-13 978 1 85578 150 4

Essays in Biochemistry volume 39: Programmed Cell Death
edited by T.G. Cotter
2003 ISBN-10 1 85578 148 4
ISBN-13 978 1 85578 148 1

Essays in Biochemistry volume 38: Proteases in Biology and Medicine
edited by N.M. Hooper
2002 ISBN-10 1 85578 147 6
ISBN-13 978 1 85578 147 4

Essays in Biochemistry volume 37: Regulation of Gene Expression
edited by K.E. Chapman and S.J. Higgins
2001 ISBN-10 1 85578 138 7
ISBN-13 978 1 85578 138 2

Essays in Biochemistry volume 36: Molecular Trafficking
edited by P. Bernstein
2000 ISBN-10 1 85578 131 X
ISBN-13 978 1 85578 131 3

Essays in Biochemistry volume 35: Molecular Motors
edited by G. Banting and S.J. Higgins
2000 ISBN-10 1 85578 103 4
ISBN-13 978 1 85578 103 0

volume 43 2007

Essays in Biochemistry

Oxygen Sensing and Hypoxia-Induced Responses

Edited by C. Peers

Portland Press

Essays in Biochemistry is published by Portland Press Ltd on behalf of the Biochemical Society

Portland Press Limited
Third Floor, Eagle House
16 Procter Street
London WC1V 6NX
U.K.
Tel.: +44 (0)20 7280 4110
Fax: +44 (0)20 7280 4169
email: editorial@portlandpress.com
www.portlandpress.com

All profits made from the sale of this publication are returned to the Biochemical Society for the promotion of the molecular life sciences.

British Library Cataloguing-in-Publication Data
A catalogue record for this book is available from the British Library
ISBN-10 1 85578 160 3
ISBN-13 978 1 85578 160 3
ISSN 0071 1365

Typeset by Aptara Inc., New Delhi, India
Printed in Great Britain by Latimer Trend and Company Ltd, Plymouth

Contents

Preface

In his excellent and highly recommended first book, the science writer Nick Lane described oxygen as the 'molecule that made the world' and illustrated just how hard it is to underestimate this gas in the development of life as we know it today. Most of us understand, and indeed take it for granted, that we inhale oxygen in order to generate energy efficiently, and this allows us to do whatever we choose to do as we go about our daily business. Healthy individuals are very unlikely, barring excursions to high altitude, to experience hypoxia (a lack of oxygen). However, numerous common cardiorespiratory diseases which unavoidably disturb the finely controlled collection, distribution and delivery of oxygen around the body can leave an individual lacking in oxygen. Such a state of hypoxia evokes a variety of striking physiological reactions which can be extremely rapid but can also have prolonged and far-reaching consequences. Such reactions illustrate just how important we regard our supply of oxygen, and they form the basis of this issue of *Essays in Biochemistry*, which attempts to bring to the reader our current understanding of the effects of hypoxia at the cellular and molecular level, in a variety of physiological and pathological situations.

Featuring heavily in several chapters is HIF (hypoxia-inducible factor), a transcription factor activated through the stabilization of its α-subunit in hypoxia, which is described accurately by Bell and Chandel as a master regulator of oxygen homoeostasis. HIF is highly influential in hypoxic remodelling of numerous cellular processes, and an excellent overview of the HIF signalling pathway is provided by Coleman and Ratcliffe. The main intracellular utilizers of oxygen, the mitochondria, are considered as oxygen-sensing organelles by Bell and Chandel, who provide evidence for the importance of oxygen-dependent generation of reactive oxygen species by mitochondria in numerous cellular processes, including HIF activation. Oxygen consumption by mitochondria can be physiologically disturbed by competition from nitric oxide, a process which is described, along with its consequences, including alteration of HIF signalling, by Galkin, Higgs and Moncada.

In reality, all cells can be said to be oxygen-sensing, since they all rely on it for oxidative phosphorylation. However, physiologically, we possess tissues which can be described as specialized oxygen-sensing tissues – tissues that generate an active reflex response to levels of hypoxia too mild to affect other tissues. The longest studied of these, the carotid body, remains something of a mystery still, but what we understand of it currently is summarized eloquently by Kumar. Likewise, the pulmonary vasculature's unique contractile response to hypoxia is discussed by Evans, and the vascular theme is extended to the systemic circulation's adaptations to hypoxia in the chapter by Paffett and Walker. In these tissues and elsewhere, ion channels, which control cell excitability among other key functions, are primary targets for modulation by hypoxia; such responses are described in detail by Kemp and Peers. Also coming to recent prominence in these oxygen-sensing tissues and elsewhere is the involvement of AMPK (AMP-activated protein kinase), which is activated by a shift in the cellular AMP/ATP ratio and, as such, is regarded as a 'metabolic fuel gauge'. This kinase is discussed in detail by Hue and Rider.

The chapter by LaManna is of particular personal interest, since it highlights an issue many researchers choose to ignore. This describes the supply and utilization of oxygen by the central nervous system, and importantly reminds us that the brain (as well as other tissues) regards normoxia as a level of oxygen that is far lower than is commonly used during experimental procedures. Indeed, the brain's own normoxia would commonly be regarded as hypoxia by many researchers.

The influence of oxygen levels in pathological situations are detailed in two further chapters. In cancer, as detailed by Brahimi-Horn and Pouysségur, HIF once again plays a fundamentally important (yet dangerous) role, controlling survival of cancerous cells in the face of extreme hypoxia, as well as promoting tumour growth through angiogenesis, and strongly influencing metastasis. Periods of hypoxia in the central nervous system also predispose individuals to dementias, the most common of which is Alzheimer's disease. Peers, Pearson and Boyle explain what we understand of this process, and how it involves dramatic remodelling of fundamental cell functions such as calcium homoeostasis.

It has been a pleasure for me to edit this issue. My thanks go to the authors, all of whom are leading experts in their respective fields: they all proved most willing to contribute, and have delivered high-quality essays with a punctuality which is as refreshing as it is unusual among the academic scientific community. My thanks also go to Roheena Anand, Michael Cunningham and especially Clare Curtis at Portland Press who have been so polite in correcting my spelling and have kept the whole process of assembling this volume running so smoothly.

Chris Peers
August 2007

Authors

Mathew Coleman is a post-doctoral researcher in Peter Ratcliffe's laboratory at the University of Oxford. He undertook his PhD at the Institute of Cancer Research, London, where he studied the role of small GTPases in the regulation of cell proliferation and apoptosis. He completed a post-doctoral position at the University of Pennsylvania, Philadelphia, U.S.A. before moving to Oxford in 2004 in order to identify novel substrates of the hypoxia-inducible factor hydroxylases. **Peter Ratcliffe** is Nuffield Professor and Head of the Department of Clinical Medicine, University of Oxford. Since 1990 his laboratory has focused on the understanding of oxygen-sensitive cell signalling. He was elected to the Fellowship of the Royal Society in 2002 and to the Membership of EMBO in 2006.

Eric Bell is a graduate student in the integrated graduate programme at Northwestern University, Chicago, U.S.A. He received his BS in biology at the University of Delaware, Delaware, U.S.A. **Navdeep S. Chandel** is an Associate Professor of Medicine and Cell/Molecular Biology at Northwestern University, Chicago, U.S.A. He received his BA in mathematics and PhD in cell physiology at the University of Chicago, U.S.A.

Alexander Galkin did his PhD thesis on the mechanism of functioning of mitochondrial Complex I in the laboratory of Andrei Vinogradov in Moscow State University and then continued to work in the field of bioenergetics. His interest in the production of reactive oxygen species and the pathological consequences of nitrosative stress began as a post-doctoral worker at Frankfurt University in the 2000s. He is currently a Senior Research Fellow at the Wolfson Institute for Biomedical Research, University College London. **Annie Higgs** has worked in the areas of inflammation and the cardiovascular actions of prostacyclin and nitric oxide. She has co-authored many reviews and edited a number of books in these research fields. She is currently a scientific co-ordinator at the Wolfson Institute for Biomedical Research, University College London. **Salvador Moncada** was Director of Research at the Wellcome Research Laboratories from 1986 until 1995. He is now Director of The Wolfson Institute for Biomedical Research at University College London. His research interests include thrombosis, atherosclerosis and inflammation, and the pharmacological actions of vasoactive substances, in particular nitric oxide and the products of arachidonic acid metabolism. He is a Fellow of the Royal Society and of the Royal College of Physicians, as well as a foreign associate of the National Academy of Sciences of the U.S.A.

Prem Kumar completed his DPhil at the University of Oxford in 1986 under the supervision of Bob Torrance and Piers Nye and completed a post-doctoral appointment at the University of Reading where he worked with Mark Hanson before taking up a Lectureship and subsequently a Readership at the University of Birmingham, U.K. He has held a Lister Fellowship and has a long-standing interest in carotid body transduction mechanisms and the reflex responses to their stimulation. He has held editorial positions on the *Journal of Physiology* and *Respiratory Physiology & Neurobiology* and is presently also a member of the Executive of The Physiological Society.

Mark Evans studied for his PhD under the supervision of Professor B.L. Ginsborg at the University of Edinburgh, on the subject of stimulus-secretion coupling in secretory epithelial cells. He remained in Edinburgh after completing his PhD in 1990 to investigate mechanisms of activation and blockage of single nicotinic acetylcholine channel currents. In 1992 he moved to St Thomas' Hospital Medical School, University of London and began to focus on the modulation by hypoxia of potassium channels in pulmonary arterial smooth muscle cells. He was awarded a Wellcome Trust Research Career Development Fellowship in 1996 to continue his work at the University of Oxford, where he developed an interest in the contribution of calcium-mobilizing pyridine nucleotides to stimulus-response coupling in pulmonary arterial smooth muscle. In 2001 he was appointed to a Lectureship at the University of St Andrews before being promoted to Reader in 2005, and broadened his studies to encompass the role of AMP-activated protein kinase in chemotransduction by hypoxia in oxygen-sensing cells.

Paul Kemp is Professor of Physiology in the School of Biosciences, Cardiff University. He obtained his DPhil at the University of Oxford in 1990 and has worked at the Universities of Dundee, Manchester, Leeds and Southern California. He has a wide range of research interests but works principally in the area of cellular and molecular physiology. His interests in oxygen sensing began in the late 1990s as a lecturer at Leeds University where he began a fruitful collaboration with Chris Peers, before moving to Cardiff in 2004. His current research focuses on the molecular mechanisms underpinning lung development, chemoreception, epithelial transport and neuronal stem cell fate determination. Presently, he is funded by the British Heart Foundation, the BBSRC (Biochemical and Biotechnology Science Research Council) and the MRC (Medical Research Council). **Chris Peers** is Professor of Cellular Physiology in the Faculty of Medicine, University of Leeds, U.K. His interests in hypoxia began as a post-doctoral worker at Oxford University in the 1980s and have developed into many aspects of his research, including the influence of hypoxia in Alzheimer's disease and remodelling of the cardiovascular system. In 2001 he was awarded the GL Brown Prize lectureship by the Physiological Society, and the following year was appointed to his chair position. He is currently Director of the Cardiovascular Research Institute at Leeds (CRISTAL; http://www.leeds.ac.uk/medhealth/light/cristal/index.htm) and also heads the Yorkshire branch of the Alzheimer's Research Trust Network.

Jayasri Nanduri, PhD, is an Associate Professor of Medicine at the University of Chicago, U.S.A. Her research interest is in protein trafficking during hypoxia with a focus on ion channel proteins especially HERG (human ether a go-go). **Nanduri R. Prabhakar**, PhD, DSc, is Professor of Medicine and Director of the Center for Systems Biology at the University of Chicago. His research interest is cellular and systemic responses to chronic continuous and chronic intermittent hypoxia.

Michael Paffett is a research associate in the Vascular Physiology Group at the University of New Mexico Health Sciences Center, Albuquerque, U.S.A. His research interests are in pulmonary endothelial cell biology, specifically how chronic hypoxia-induced pulmonary hypertension alters calcium influx and sequestration pathways. **Benjimen Walker** is a Professor of Cell Biology and Physiology and heads the Vascular Physiology Group at the University of New Mexico Health Sciences Center, Albuquerque, U.S.A. His research centres on the effects of chronic hypoxia on vascular regulation.

Louis Hue is Full Professor of Biochemistry at the Université catholique de Louvain, Medical Faculty, Brussels, Belgium and Head of a Research Unit at the Christian de Duve ICP (Institute of Cellular Pathology), also in Brussels. He studied medicine and obtained his MD degree in 1967. He spent the next 12 years in the laboratory of Professor H.G. Hers at the ICP where he mainly studied the role of futile cycles in the regulation of carbohydrate metabolism. This led to his PhD degree (1979) and to the discovery of fructose 2,6-bisphosphate with E. Van Schaftingen and H.G. Hers in 1980. He spent 1 year of post-doctoral research in the laboratory of J.H. Exton at Vanderbilt University, Nashville, U.S.A. After his return to the ICP, he started his research group and was Head of the Department of Biochemistry. His main interest is still in the control of carbohydrate metabolism, with special emphasis on insulin-activated protein kinases and AMP-activated protein kinase. **Mark Rider** is lecturer in Biochemistry at the Université catholique de Louvain Faculty of Medicine, Brussels, Belgium. He graduated with a BSc in Biochemistry at Bristol University and obtained a PhD in Biochemistry at University College London in the laboratory of David Saggerson in 1983. Thereafter he joined the group of Louis Hue in Brussels to study the role of fructose 2,6-bisphosphate in the control of glycolysis in mammalian tissues. His interest in protein phosphorylation is long-standing, particularly in the control of metabolism, protein synthesis and muscle contraction.

Joseph C. LaManna, PhD, is Professor and Chair of Anatomy in the School of Medicine at CWRU (Case Western Reserve University), Cleveland, U.S.A., where he has spent the last 25 years studying how the brain uses oxygen and glucose to supply metabolic energy to support neuronal work. He received his doctorate from Duke University, Durham, U.S.A. in physiology and pharmacology and, prior to moving to CWRU, was at the University of Miami in Florida.

Hugh A. Pearson is a Senior Lecturer in the Institute of Membrane and Systems Biology, Faculty of Biological Sciences, University of Leeds, U.K. He has maintained an interest in neuronal ion channel function since his post-doctoral investigations at the University of London. Since the mid-1990s he has become interested in the role of ion channels in dementia and has studied the effects of amyloid proteins in primary neurons from the central nervous system. He is also interested in a physiological role for this peptide and the possible deleterious consequences of blocking its function as a treatment for Alzheimer's disease. **John P. Boyle** is a Senior University Cardiovascular Research Fellow in CRISTAL, (the Cardiovascular Research Institute at Leeds) in the University of Leeds, U.K. Prior to this appointment his interests were the control of blood flow to the brain; in particular the control of cerebral arterioles and of the cerebral capillary endothelial cells of the blood–brain barrier. Since joining with Chris Peers in 2001 he has continued his work on the control of cellular calcium, now additionally in astrocytes and neurons. His current scientific interests are in the effects of hypoxia on calcium homoeostasis and ion channel physiology and the possible role this plays in the development of Alzheimer's disease.

Christiane Brahimi-Horn is a Senior Research Scientist employed by the French research agency INSERM. She has been working in the CNRS UMR6543 laboratory in Nice on the hypoxia-inducible factor pathway for the last 6 years. She has previous experience in the areas of glycoconjugated biochemistry, gene therapy and biotechnology obtained in France (INSERM and CNRS), Taiwan [DCB (Development Center for Biotechnology)] and Australia [CSIRO (Commonwealth Scientific and Research Organization)]. She obtained a BSc degree with Honours from Monash University, Victoria, Australia and a PhD from the University of Paris VI in France in 1981. **Jacques Pouysségur** is the Director of the Institute of Signaling, Development, Biology and Cancer Research, a CNRS Unit housing 13 group leaders (http://www.unice.fr/isdbc/). He has previous experience in anaerobic glycolysis, pH regulation, Na^+/H^+ exchange and mitogen-activated protein kinase signalling in the context of growth control in mammalian cells. His group is now involved in hypoxia signalling related to tumour angiogenesis. He graduated from an Engineering School in Biochemistry and obtained his PhD in 1972 from the University of Lyon in France. He spent 2 years as a post-doctoral scientist at the National Cancer Institute of NIH (National Institutes of Health) and a 1 year sabbatical at MIT (Massachusetts Institute of Technology). He is a member of the French and European Academy of Science.

Abbreviations

2-OG	2-oxoglutarate
4E-BP1	4E-binding protein
Aβ	amyloid β peptide
ACC	acetyl-CoA carboxylase
AD	Alzheimer's disease
ADAM	**A** **D**isintegrin **A**nd **M**etalloprotease
AE2	anion exchanger isoform 2
AICA	5-amino-4-imidazole-carboxamide
AMF	autocrine motility factor
AMPK	AMP-activated protein kinase
Ang-2	angiopoietin-2
AP-1	activator protein-1
APP	amyloid precursor protein
ARD	ankyrin repeat domain
ARK	AMPK-related kinase
ARNT	aryl hydrocarbon receptor nuclear translocator
AS160	Akt substrate 160
BACE-1	β-amyloid cleavage enzyme
bHLH	basic helix–loop–helix
BNIP3	Bcl-2/adenovirus E1B 19 kDa interacting protein 3
BOLD	blood oxygen level dependent
B*r*	bilirubin
B*v*	biliverdin
B*v*R	B*v* reductase
CA	carbonic anhydrase
CAD	C-terminal transactivation domain
cADPR	cyclic adenosine diphosphate-ribose
CaMK	calcium/calmodulin-dependent protein kinase II
CaMKK	calmodulin-dependent protein kinase kinase
CBP	cAMP-response-element-binding protein-binding protein
CBS	cystathionine β-synthase
CH	chronic hypoxia
CMRglu	cerebral metabolic rate for glucose
$CMRO_2$	cerebral metabolic rate for oxygen
COPD	chronic obstructive pulmonary disease
COX	cyclo-oxygenase
CREB	cAMP-response-element-binding protein

DSBH	double-stranded β-helix
ECM	extracellular matrix
EDRF	endothelium-derived relaxing factor
eEF2	eukaryotic elongation factor 2
eIF	eukaryotic initiation factor
E_m	resting membrane potential
eNOS	endothelial nitric oxide synthase
EPO	erythropoietin
ERK1/2	extracellular-signal-regulated kinase 1/2
ET-1	endothelin-1
FH	fumarate hydratase
FIH	factor inhibiting HIF
F_iO_2	fraction inspired oxygen
fMRI	functional magnetic resonance imaging
FRET	fluorescence resonance energy transfer
G6PD	glucose-6-phosphate dehydrogenase
GAP	GTPase-activating protein
GDI	guanine nucleotide dissociation inhibitor
GEF	guanine-nucleotide-exchange factor
GLUT	glucose transporter
GPX	glutathione peroxidase
HEK	human embryonic kidney
HET	heterozygous
HIF	hypoxia-inducible factor
HMG-CoA	hydroxymethylglutaryl-CoA
HO	haemoxygenase
HPH	hypoxic pulmonary hypertension
HPV	hypoxic pulmonary vasoconstriction
HRE	hypoxia-responsive element
IGF	insulin-like growth factor
IH	intermittent hypoxia
IκB	inhibitory κB
IKK	IκB kinase
IL-8	interleukin-8
iNOS	inducible nitric oxide synthase
IP_3	inositol trisphosphate
Kv	voltage-sensitive potassium
LDH-A	lactate dehydrogenase A
LTF	long-term facilitation
MAPK	mitogen-activated protein kinase
MCT	monocarboxylate transporter
MLC	myosin light chain
MLCK	myosin light chain kinase

MLCP	myosin light chain phosphatase
MnTMPyP	manganese (III) tetrakis-(1-methyl-4-pyridyl)-porphyrin pentachloride
mtNOS	mitochondrial nitric oxide synthase
mTOR	mammalian target of rapamycin
NAD	N-terminal transactivation domain
NF-κB	nuclear factor κB
NHE1	Na^+/H^+ exchanger
NIP3L	Nip-3 like
NIX	protein X
NLS	nuclear localization signal
nNOS	neuronal nitric oxide synthase
NO	nitric oxide
NOS	nitric oxide synthase
Nox	NADPH oxidase
Noxa1	NADPH oxidase activator 1
Noxo1	Nox organizer 1
NTS	nucleus tractus solitarius
ODDD	oxygen-dependent degradation domain
$ONOO^-$	peroxynitrite
OS-9	amplified in osteosarcoma 9
*p*50	oxygen partial pressure producing 50% saturation
p70S6K	p70 ribosomal protein S6 kinase
PAS	Per-Arnt-Sim
PC12	pheochromocytoma-12
PCO_2	partial pressure of carbon dioxide
PDH	pyruvate dehydrogenase
PDK1	pyruvate dehydrogenase kinase 1
PET	positron emission tomography
PFK-2	6-phosphofructo-2-kinase
PGE_2	prostaglandin E_2
PGI	phosphoglucose isomerase
PGM	phosphoglycerate mutase
PHD	prolyl hydroxylase domain
pHe	extracellular pH
pHi	intracellular pH
phox	phagocytic oxidase
PI3K	phosphoinositide 3-kinase
PKA	cyclic AMP-dependent protein kinase
PKB	protein kinase B
PO_2	partial pressure of oxygen
PTEN	phosphatase and tensin homologue deleted on chromosome 10

RCC	renal cell carcinoma
Rheb	Ras homologue enriched in brain
RNAi	RNA interference
ROS	reactive oxygen species
SCO2	cytochrome *c* oxidase 2
SDH	succinate dehydrogenase
sGC	soluble guanylate cyclase
SOD	superoxide dismutase
SR	sarcoplasmic reticulum
TAD	transcriptional activation domain
TAK	transforming growth factor-β-activated kinase
TCA	tricarboxylic acid
TGF-α	transforming growth factor-α
TGF-β	transforming growth factor-β
TH	tyrosine hydroxylase
TIGAR	TP53-induced glycolysis and apoptosis regulator
TNFα	tumour necrosis factor α
TORC2	transducer of regulated CREB activity 2
TRP	transient receptor potential
TSC	tuberous sclerosis complex
VEGF	vascular endothelial growth factor
VGCC	voltage-gated calcium channel
VHL	von Hippel-Lindau tumour-suppressor
VSM	vascular smooth muscle
WT	wild-type

1

Oxygen sensing and hypoxia-induced responses

Mathew L. Coleman and Peter J. Ratcliffe[1]

Henry Wellcome Building for Molecular Physiology, University of Oxford, Roosevelt Drive, Oxford OX3 7BN, U.K.

Abstract

Low cellular oxygenation (hypoxia) represents a significant threat to the viability of affected tissues. Multicellular organisms have evolved a highly conserved signalling pathway that directs many of the changes in gene expression that underpin physiological oxygen homoeostasis. Oxygen-sensing enzymes in this pathway control the activity of the HIF (hypoxia-inducible factor) transcription factor by the direct incorporation of molecular oxygen into the post-translational hydroxylation of specific residues. This represents the canonical hypoxia signalling pathway which regulates a plethora of genes involved in adaptation to hypoxia. The HIF hydroxylases have been identified in other biological contexts, consistent with the possibility that they have other substrates. Furthermore, several intracellular proteins have been demonstrated, directly or indirectly, to be hydroxylated, although the protein hydroxylases responsible have yet to be identified. This chapter will summarize what is currently known about the canonical HIF hydroxylase signalling pathway and will speculate on the existence of other oxygen-sensing enzymes and the role they may play in signalling hypoxia through other pathways.

Introduction

Molecular oxygen acts as the terminal electron acceptor in mitochondrial oxidative phosphorylation. As such it is required for efficient ATP production,

[1]*To whom correspondence should be addressed (email pjr@well.ox.ac.uk).*

which is essential for most biological processes. Metazoans have evolved specialized systems that respond to oxygen starvation by altering cell metabolism and survival, and rescuing oxygen delivery to the affected tissue. A key player mediating hypoxia-induced responses is the HIF (hypoxia-inducible factor) transcription factor complex (reviewed in [1–8]). Oxygen-dependent regulation of HIF activity controls an array of hypoxia-inducible genes that together programme the cell to respond to changes in oxygen availability.

HIF is a heterodimer consisting of a constitutively expressed β-subunit and an α-subunit which is induced when oxygen levels are low (Figure 1) [4,6,8]. HIF-α and HIF-β are members of the PAS (Per-Arnt-Sim) family of bHLH (basic helix-loop-helix) proteins (Figure 2), each of which are expressed as three proteins from independent genes. Of the α-subunits, HIF-1α and HIF-2α display the greatest similarity both in terms of sequence homology,

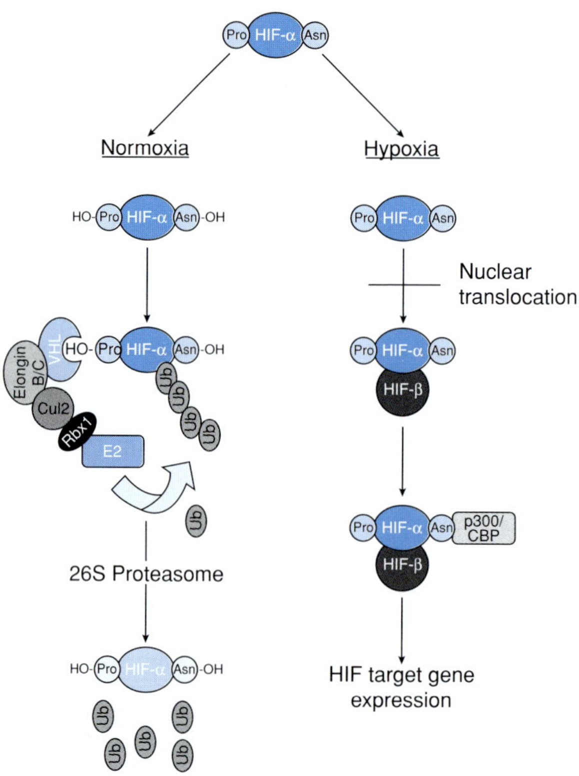

Figure 1. Regulation of HIF activity by prolyl and asparaginyl hydroxylation
In the presence of oxygen (normoxia), asparaginyl hydroxylation (OH) of the HIF-α CAD inactivates HIF transcriptional activity. Prolyl hydroxylation creates a binding site for the VHL E3 ubiquitin (Ub) ligase complex, which consists of VHL, elonginB/C, Cul2 and Rbx1. Subsequent transfer of Ub from E1 and E2 enzymes to HIF-α signals its rapid degradation by the 20S proteasome. When oxygen is limiting (hypoxia), hydroxylation is prevented. HIF-α escapes VHL capture and is stabilized. Following nuclear translocation and binding to HIF-β and p300/CBP, the HIF complex is transcriptionally competent, which results in the expression of genes involved in adaptation to hypoxia.

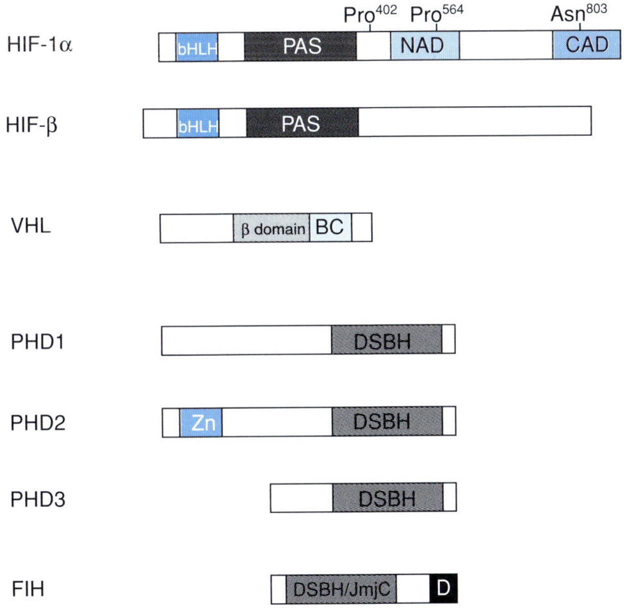

Figure 2. Components of the HIF oxygen signalling pathway
bHLH, basic helix-loop-helix domain involved in dimerization and DNA binding; PAS, Per/Arnt/Sim domain; NAD, N-terminal transactivation domain; CAD, C-terminal transactivation domain; β-domain, motif that binds hydroxy-Pro HIF-α; BC, BC box (recruits the elongin B/C complex); DSBH, double-stranded β-helix motif (catalytic core of 2-oxoglutarate/Fe(II)-dependent dioxygenases); Zn, zinc-finger like motif; JmjC, DSBH motif common to a subset of 2-OG-/Fe(II)-dependent enzymes; D, dimerization domain.

and function. HIF-3α is more divergent, and its regulation and function are less clearly defined. Under hypoxic conditions, HIF-1α and HIF-2α protein levels increase dramatically. This response is followed by nuclear translocation, association with HIF-β, recruitment of co-activators and activation of transcription, with at least some of these processes constituting independent points of control by oxygen (Figure 1) [8].

HIF target genes and adaption to hypoxia

Genes that are direct HIF transcriptional targets contain a G/ACGTG motif termed the HRE (hypoxia-responsive element); such HREs being found in genes that regulate both systemic and cellular responses to hypoxia. The HIF system was discovered in studies of the strikingly hypoxia-responsive EPO (erythropoietin) gene, as a DNA-binding complex mediating the activity of an HRE in the *EPO* 3′-flanking region [6]. In hypoxia, EPO production increases several hundred fold and enhances the oxygen-carrying capacity of the blood by stimulating red blood cell production.

It is now recognized that HIF also promotes delivery of oxygen to hypoxic tissues by stimulating angiogenesis via the upregulation of growth factors such as the VEGF (vascular endothelial growth factor) family and its receptors. Other systemic responses mediated by HIF include the regulation of vascular tone through hypoxic induction of genes encoding vasomotor regulators such as nitric oxide synthases, haem oxygenase, endothelin-1 and adrenomedullin [4].

Cellular responses to hypoxia controlled by HIF include changes in energy metabolism, cell growth, survival and migration. HIF directs the co-ordinate regulation of many genes in these pathways. For instance HREs are present in the genes encoding glucose transporters and glycolytic enzymes, as well as regulators of aerobic and anaerobic metabolism [5,8]. Together they switch metabolism from oxidative phosphorylation to anaerobic glycolysis in order to maintain ATP production during hypoxia.

Another critical response to chronic hypoxia is cell death, a process in which HIF has also been implicated. Paradoxically, HIF also up-regulates genes involved in cell survival, suggesting that in some circumstances cells may be subjected to conflicting HIF-dependent signals. The outcome may be influenced by the severity and duration of the oxygen starvation or associated stresses such as acidosis. Hypoxic cell death has been associated with transcriptional activation of the Bcl-2 family member, BNIP3 (Bcl-2/adenovirus E1B 19kDa interacting protein 3) [1]. In contrast, VEGF and EPO can have positive effects on cell survival. In addition, IGF (insulin-like growth factor) 2 and TGFα (transforming growth factor α) are HIF targets that activate intracellular survival pathways [7].

HIF target genes have also been implicated in the control of cell migration and invasion [3]. HIF-mediated induction of the receptor tyrosine kinase, Met, promotes cell motility through increased activity of the hepatocyte growth factor signalling pathway. Invasion of cells through the basement membrane is in turn promoted by repression of E-cadherin, an important component of adherens junctions. Such pathways are co-opted by hypoxic cells, and by tumour cells that constitutively express HIF. For instance, HIF-dependent induction of the chemokine receptor CXCR4 promotes the chemotaxis of renal cell carcinoma cells.

Overall, recent microarray analyses have demonstrated that hundreds or even thousands of genes respond to hypoxia in any given type of cell. Somewhat surprisingly, genetic manipulation of the HIF system has indicated that, at least under standard tissue culture conditions, substantially more than 50% of these genes respond directly or indirectly to HIF [9].

Post-translation hydroxylation of HIF-α

So far two steps in the regulation of HIF by oxygen have been clearly defined. These involve regulation of both the abundance and the transcriptional activity of HIF-α subunits by different types of post-translation hydroxylation.

Prolyl hydroxylation

In the presence of oxygen, cellular levels of HIF-α proteins are suppressed by rapid constitutive degradation via the ubiquitin–proteasome pathway. Two independent domains in HIF-1α and HIF-2α mediate this oxygen-dependent degradation, both positioned around conserved proline residues, Pro^{402} and Pro^{564} in HIF-1α (Figure 2) [5]. Hydroxylation of either of these residues creates a binding site for the VHL (von Hippel-Lindau tumour-suppressor) protein (Figure 2), the substrate-docking site of an E3 ubiquitin ligase complex containing Elongin B/C, Cul2 and Rbx1 proteins (Figure 1) [2]. Subsequent ubiquitylation of prolyl-hydroxylated HIF-α signals its rapid degradation by the 26S proteasome.

HIF-α prolyl hydroxylation is catalysed by a family of PHD (prolyl hydroxylase domain) enzymes (PHD1, 2 and 3) (Figure 2) that belong to a larger family of enzymes which require oxygen, 2-OG (2-oxoglutarate) and Fe(II) for activity [5]. Members of this family are characterized by a catalytic core of eight β-strands that fold into a DSBH (double-stranded β-helix) motif. This structural arrangement brings together ligands involved in Fe(II) and 2-OG binding at the PHD active site, which is highly conserved in all three enzymes, both in sequence and predicted tertiary structure. Despite this, there are differences in substrate specificity between the PHDs. PHD1 and 2 hydroxylate Pro^{402} and Pro^{564} residues in HIF-1α and the equivalent residues in HIF-2α, whereas PHD3 only targets Pro^{564} or the equivalent [5]. The PHD2 crystal structure suggests that substrate specificity may be partly explained by variation in the β2-β3 finger motif that is distal to the iron centre [10]. However, there are other differences between the PHDs that could contribute to selectivity.

PHD1 and PHD2 have a similar domain organization with a C-terminal catalytic domain and an N-terminal extension (Figure 2). In the case of PHD2 the N-terminus contains a zinc-finger-like domain that may be involved in autoinhibition of the enzyme [11]. PHD3 is significantly smaller in size because it lacks the N-terminal extension present in the other PHDs. In addition to specific activity, differences in subcellular localization, interactions with other proteins, and relative abundance of the proteins in cells also influence the effective contribution of different HIF-α isoforms to degradation *in vivo* (reviewed in [2,12]).

Although the three PHDs isoforms are closely related, they do not appear to be redundant. PHD2 is responsible for the majority of prolyl hydroxylase activity under normoxia; and PHD2 RNAi (RNA interference) alone is sufficient to induce HIF-1α protein expression and HIF target gene expression [13]. Further evidence for the importance of PHD2 has come from genetic analyses of inherited erythrocytosis (an increased number of red blood cells) A P317R PHD2 mutation associated with familial erythrocytosis significantly reduced PHD2 activity [14]. Structural studies rationalize this effect, and indicate that P317R most likely affects both Fe(II) and substrate binding [10]. Furthermore, knockout mouse studies have shown that whereas PHD2 null mice are embryonically lethal due to defects in heart and placental development [15], PHD1 and PHD3 null mice are viable. It is likely that these differences are due in part to the more widespread expression of PHD2 relative to that of PHD1 and PHD3 (see below) [16].

Regulation of PHD activity

The physiological challenge of oxygen homoeostasis requires a system capable of flexible responses over a wider range of oxygen tensions. In this respect it is interesting that the total level of PHD activity in a cell is regulated by a number of factors in addition to oxygen availability, including abundance of the PHD enzymes, cofactor availability, and interaction with other proteins.

Abundance of PHDs

The expression level of each PHD is increased by specific signals (see below), including hypoxia. Indeed, PHD2 and PHD3 appear to be direct HIF targets. This provides a negative feedback loop that, in response to temporal changes in oxygenation, may serve to limit HIF-α induction in hypoxia or accentuate the response to reoxygenation. At steady-state it would be predicted to provide a 'range-extending' function by adjusting hydroxylase activity in accordance with available oxygen (see below) (reviewed in [12]). Interestingly, whereas PHD2 appears to be ubiquitously expressed, other isoforms show marked tissue specificity; for instance PHD1 is highly expressed in testes whereas PHD3 is most abundant in the heart [17]. Nevertheless the effects of these differences and whether they alter the oxygen-sensing range in these tissues is unclear.

There are several mechanisms by which PHD expression is regulated, including transcription, translational initiation, mRNA splicing and protein stability. Thus PHD1 is expressed in a variety of cells as two proteins arising from distinct sites of translational initiation [18]. Alternatively spliced transcripts of PHD2 and 3 have also been described, most of which are enzymatically inactive [19,20]. Interestingly, some of these other PHD forms have the potential to exert dominant-negative effects. PHD3 is a highly inducible short-lived protein that is targeted for proteasomal degradation by Siah E3 ubiquitin ligases [21]. The importance of this control has been illustrated in mice bearing targeted inactivation of the Siah2 gene where endogenous PHD3 is elevated and the EPO response to hypoxia is blunted. Though the Siah ubiquitin ligase can also interact with PHD1, its physiological role is less clear in this context and, at least in cell culture, PHD1 levels are not significantly affected by genetic manipulation of Siah1 and 2 [18].

Interacting proteins

A number of binding partners have been identified for the PHDs, including those that are likely to promote protein folding and others that modulate activity towards HIF-α. Thus subunits of the cytosolic chaperone TriC were identified by MS in association with over-expressed PHD3 but not other PHDs [22]. OS-9 (amplified in osteosarcoma 9) was identified in a yeast-2-hybrid screen using the C-terminus of HIF-1α as bait, and subsequently shown to interact with PHD2 and PHD3 [23]. OS-9 overexpression may promote HIF-α hydroxylation and degradation under certain conditions, possibly by the formation of a HIF-α–PHD–OS-9 ternary complex. Similarly, PHD3 is reported to interact with the WD-repeat (tryptophan-aspartate repeat) protein Morg-1

[MAPK (mitogen-activated protein kinase) organizer 1], an association that is postulated to suppress HIF-α by promoting PHD3-mediated hydroxylation [24]. It is still unclear what the physiological roles of these processes are, although it is likely that together they contribute to a system that tightly controls HIF-α degradation in a context- and cell type-dependent manner.

Cofactor availability

The PHDs belong to a large family of 2-OG/Fe(II)-dependent dioxygenases that catalyse oxidative modifications (reviewed in [25]). Oxygen is split into two atoms during the catalytic cycle. One is incorporated into the prime substrate (e.g. HIF-α) while the other is coupled to the oxidative decarboxylation of 2-OG to yield succinate and carbon dioxide (Figure 3). Fe(II) co-ordinates 2-OG and oxygen towards the prime substrate and forms the reactive intermediate that drives the oxidative process. It is becoming clear that the availability of essential cofactors such as oxygen (see below), Fe(II) and 2-OG are important determinants of PHD activity (Figure 3), and therefore HIF-α expression.

Recent studies suggest that certain TCA (tricarboxylic acid) cycle intermediates can act as natural PHD inhibitors by competing with 2-OG (reviewed in [26]), and it has been proposed that consequential activation of HIF promotes the growth of tumours associated with genetic defects in the TCA cycle. Thus SDH (succinate dehydrogenase) converts succinate into fumarate, and its inactivation is associated with a predisposition to paraganglioma and phaeochromocytoma. Similarly, fumarate is converted into malate by FH (fumarate hydratase), and inactivating mutations of the *FH* gene predisposes to a variety of tumour types. Tumours harbouring *SDH* or *FH* mutations are generally highly vascularized and express elevated levels of HIF-α. Thus an oncogenic pathway based on accumulation of succinate or fumarate, competitive inhibition of PHD activity [26], and increased activity of HIF (or other) PHD substrates has been proposed.

The importance of Fe(II) for PHD activity is well demonstrated by the classical characteristics of induction of the HIF system by either divalent cations, which substitute for Fe(II) in the active site, or by compounds that chelate the cellular iron pool [5]. However, it is becoming clear that other more physiological stimuli may also affect PHD activity by actions on Fe(II) at the active site. Rapidly growing cells are often iron deficient and supplementation with iron and/or ascorbate is sufficient to down-regulate normoxic HIF α, most likely by enhancing PHD activity. The exact role of ascorbate in promoting PHD activity remains unclear, but may involve the reduction of Fe(III) at the active site or an action to increase the availability of Fe(II) [5]. Interestingly, ascorbate is sufficient to rescue PHD activity in cells that overproduce ROS (reactive oxygen species). For example, JunD regulates genes involved in antioxidant defence and JunD-deficient cells manifest activation of HIF in association with accumulation of hydrogen peroxide and ROS. It has been proposed

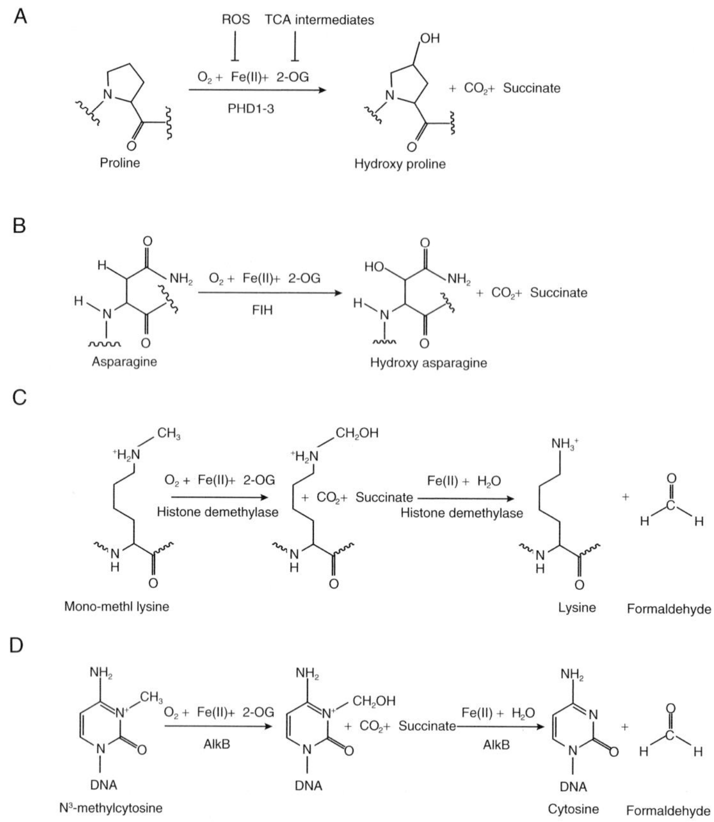

Figure 3. Reactions catalysed by HIF-α hydroxylases and other 2-OG/Fe(II)-dependent dioxygenases
(A) PHD1–3 hydroxylate Pro^{402} and Pro^{564} of HIF-1α, and the equivalent residues of HIF-2α. *Trans*-4-hydroxylation of these residues is dependent on dioxygen, Fe(II) and 2-OG and is compromised by hypoxia, ROS and TCA intermediates respectively. The oxidative decarboxylation of 2-OG forms carbon dioxide and succinate, with one oxygen atom from dioxygen incorporated into the prime substrate, and the other into succinate. **(B)** FIH hydroxylates the *pro-S* position of the β-carbon of HIF-1α Asn^{803}, and the equivalent position of HIF-2α. **(C)** JmjC domain-containing histone demethylases catalyse hydroxylation of the methylamine group of methylated lysine residues. In the presence of dioxygen, Fe(II) and 2-OG these enzymes produce hydroxymethyl-lysine, succinate and carbon dioxide. The unstable hydroxymethyl group is spontaneously lost as formaldehyde, leaving a methyl group. **(D)** Human homologues of *E. Coli.* AlkB repair alkylated DNA by catalysing the hydroxylation of 3-methylcytosine (and 1-methyladenine). Hydroxylation of the methyl group produces a hydroxy intermediate that decomposes to the repaired base plus formaldehyde.

that the increase in ROS leads to the oxidation of Fe(II) to Fe(III) at the PHD2 catalytic centre thus inducing HIF-α even in the presence of oxygen [27].

Other investigators have proposed that production of mitochondrial ROS plays a role in PHD inactivation under hypoxia. In support of this, mitochondrial defects that reduce ROS production are associated with reduced induction of HIF in moderate hypoxia (reviewed in [28]). However an alternative explanation is that these defects simply reduce the severity of intracellular hypoxia by reducing oxygen consumption. It should also be noted that the

PHDs are able to rapidly hydroxylate HIF-α upon reoxygenation, suggesting that either they are not in fact damaged in hypoxia, or that they are very rapidly repaired in the presence of oxygen.

Asparaginyl hydroxylation

In addition to the oxygen-dependent degradation domains, the HIF-α C-terminus mediates a second oxygen sensitivity control of HIF activity. This region contains the CAD (C-terminal transactivation domain) (Figure 2) and is responsible for recruiting the p300/CBP (cAMP-response-element-binding protein-binding protein) transcriptional co-activator to HIF target gene promoters (reviewed in [2]). HIF-α CAD activity is suppressed in normoxia due to the incorporation of oxygen into a conserved asparagine residue by FIH (factor inhibiting HIF) which, like the PHDs, is also a 2-OG/Fe(II)-dependent dioxygenase (Figure 2) [5]. FIH-mediated HIF-α CAD asparaginyl hydroxylation prevents the interaction with the p300 CH1 domain. Unlike the PHDs, FIH appears to be abundant and ubiquitously expressed, with protein levels relatively constant between cell types and conditions [29]. Although FIH targets the CADs of both HIF-1α and HIF-2α, it does so to differing extents. HIF-2α is less responsive to FIH than HIF-1α due to a conserved amino acid substitution in HIF-2α that reduces the efficiency of asparagine hydroxylation [30].

Genetic analyses confirm that FIH is an important component of the oxygen dependent suppression of HIF; FIH RNAi is sufficient to up regulate the HIF target genes GLUT1 (glucose tranporter 1) and CA9 (carbonic anhydrase 9) in a variety of cell lines under normoxia [29]. However, neither PHD2 nor FIH RNAi elevates HIF target genes to the same extent as that observed under hypoxia. However, the combined effect of FIH and PHD2 RNAi is sufficient to induce maximal HIF activity at least in certain cells, suggesting that these enzymes are the dominant HIF hydroxylases [31]. These studies also revealed that FIH differentially regulates specific HIF target genes, perhaps reflecting target gene specific differences in the use of either of two transactivation domains that are (CAD) or are not [NAD (N-terminal transactivation domain)] affected by asparaginyl hydroxylation (Figure 2).

HIF hydroxylases as 'oxygen sensors'

Since molecular oxygen is absolutely required as co-substrate for the HIF hydroxylation reactions catalysed by both the PHDs and FIH, these enzymes are well placed to mediate an oxygen-sensing function. Nevertheless certain conditions must be fulfilled for a sensing role in physiological oxygen homoeostasis. First, the affinity for oxygen must be lower than the physiological concentration of oxygen; a drop in oxygen concentration would therefore lead to a concomitant decrease in hydroxylase activity, with a consequent increase in HIF expression. Secondly the rate of HIF hydroxylation must be rate-limiting for the overall inactivation process [12].

Several studies have sought to measure reaction kinetics of the HIF hydroxylases as a function of oxygen concentration. Initial *in vitro* studies using recombinant PHD enzymes and short HIF-1α peptides suggested a very high K_m for oxygen of approximately 250 μM. Though these early results suggested an unusually low affinity for oxygen, other work suggests that the true affinity of the PHDs for oxygen *in vivo* may be substantially higher. Thus similar experiments using a longer HIF-1α polypeptide have yielded substantially lower apparent K_m values for oxygen [32,33]. Though the order of co-substrate and substrate-binding has not yet been defined for the HIF hydroxylases, these results are compatible with studies of other enzymes in this class that indicate that it is the enzyme–substrate complex that binds molecular oxygen, allowing both components to influence affinity. Though the affinity of FIH for oxygen has been reported to be higher than the PHDs at approx. 90 μM, these results were also obtained with short peptide substrates that may behave differently from the intact HIF-α protein. Thus given the increased oxygen affinity that is observed when longer (and more physiological) HIF-α polypeptides are used in kinetic studies, it is not yet clear that the HIF hydroxylases have kinetic properties that are specialized for an oxygen-sensing role. Nevertheless given that physiological tissue oxygen concentrations are likely to be in the range of 10–30 μM, even much lower K_m values would be quite compatible with an oxygen-sensing function provided that the overall rates of HIF hydroxylation were limiting for HIF inactivation. That this is indeed the case is supported by the observation that modest changes in enzyme abundance affect the oxygen dependence of HIF activation. Thus it is important to recognize that the actual oxygen-dependence of HIF regulation is dependent on the characteristics of the system as a whole and will be determined by all the factors (see above) that influence the rate of HIF hydroxylation, as well as the affinity for molecular oxygen itself.

HIF hydroxylase substrates

Since the discovery of oxygen-sensitive signalling through protein hydroxylation in the context of HIF, attention has focused on whether a similar mechanism might operate on other systems. For instance the HIF hydroxylases themselves might have other substrates, or other enzymes in this family might regulate different systems in an oxygen-sensitive manner.

Several lines of evidence have suggested that the PHDs might have other substrates, though clear proof is lacking at the present time. First, the encoding genes have been identified in screens that are not readily connected with known functions of HIF. Thus PHD1 has been identified as a target of the oestrogen receptor and promotes the colony growth of breast cancer cells [34]. The *Drosophila* homologue of PHD2 has been implicated downstream of cyclin D/Cdk4 in a pathway that regulates cell growth [35]. PHD3 was independently cloned as a transcript induced by the differentiation of smooth muscle cells, by the activation of p53, and following nerve growth factor withdrawal in sympathetic neurones (reviewed in [5]). Further studies have indicated that PHD3 expression is necessary and sufficient for neuronal apoptosis under these conditions [36,37], and that the property is oxygen-sensitive and dependent on hydroxylase activity [36]. However, as yet it is unclear whether any of these activities derive from enzymatic activities on substrates other than HIF [36,38].

Other transcription factors are responsive to hypoxia; NF-κB (nuclear factor κB) signalling being one which has attracted considerable interest. However, compared with the HIF system, regulation by hypoxia is less robust and the pathways remain poorly understood. Cummins et al. [39] have recently suggested that NF-κB activation in hypoxic cells involves the inhibition of PHD1, which they propose normally suppresses IKK (IκB kinase) activity by hydroxylating IKKβ Pro^{191}. PHD1 overexpression was sufficient to inhibit TNFα-induced NF-κB activity, although the effect of a hydroxylation-defective PHD1 mutant was not reported, and no direct evidence for IKK hydroxylation was provided.

In contrast with current uncertainties about the nature of other substrates for the PHDs, clear evidence of alternative substrates for the asparaginyl hydroxylase FIH has been obtained. Initially a yeast-two-hybrid screen identified members of the IκB family as novel FIH-interacting proteins [40]. Conserved asparagine residues in the ARDs (ankyrin repeat domains) of p105 and IκBα are FIH substrates. Unlike HIF-α however, asparaginyl hydroxylation of IκBα does not appear to affect its activity dramatically. Recent studies on FIH-mediated hydroxylation of analogous asparagine residues in the Notch transmembrane receptor have shown that, like IκBα, the activity of the canonical Notch signalling pathway does not appear to be regulated by FIH. Nevertheless given the prevalence of the ARD domain in the mammalian proteome, these results strongly suggest that protein hydroxylation will turn out to be a common post-translational modification. Interestingly the affinity of FIH for ARDs is much higher than that for HIF so that these proteins are effective competitive inhibitors of FIH-mediated HIF-α CAD hydroxylation thereby providing an additional mechanism of modulating HIF signalling (M.L. Coleman, unpublished work).

In a number of other systems indirect evidence for protein hydroxylation or action of a 2-OG oxygenase has suggested existence of hydroxylase pathways but the enzyme(s) involved are unknown (reviewed in [5]). Thus the hyperphosphorylated form of the Rpb1 subunit of RNA polymerase II has been proposed to be prolyl hydroxylated and targeted for proteasomal degradation by VHL. Iron-regulatory protein 2 is involved in cellular iron sensing and, based on regulation by inhibitors of 2-OG/Fe(II)-dependent enzymes, has also been postulated to be a target of post translational hydroxylation. Hypoxia leads to an increase in the synthesis of phosphatidic acid by diacylglycerol kinase, and again regulation by inhibitors of this family of enzymes has suggested the involvement of an as yet unknown 2-OG oxygenase.

2-OG dioxygenases as general oxygen sensors

Sequence analyses predict that the human genome encodes over 60 2-OG/Fe(II)-dependent dioxygenases [25], most with unknown functions. As discussed above, since it is unlikely that the HIF hydroxylases require special biochemical properties to function as oxygen sensors, other family members might also signal oxygen availability if appropriately coupled to a regulatory process. Since these enzymes are not all protein hydroxylases and in fact oxidize a variety of substrates they might signal oxygen availability to a variety of cellular processes.

For instance, DNA is damaged by a variety of alkylating agents that are toxic and mutagenic. Human homologues of *Escherichia coli* AlkB are 2-OG/Fe(II)-dependent dioxygenases that repair DNA methylated at 1-methyladenine and 3-methylcytosine (reviewed in [41]). Hydroxylation of the methyl group produces a hydroxyl intermediate that decomposes to the repaired base plus formaldehyde (Figure 3). This class of enzymes may therefore represent a novel interface between hypoxia and DNA repair.

Post-translational modifications of histones (H) are important regulators of gene expression. Methyltransferases catalyse the methylation of H3 and H4 at specific lysine residues; the site of modification determines whether methylation results in transcriptional activation or silencing. Recent studies suggest that 2-OG/Fe(II)-dependent dioxygenases with similarity to FIH can reverse this process (reviewed in [42]). These enzymes regulate gene expression and the proliferation of tumour cells. They possess what has previously been termed a JmjC (Jumonji C) domain, which is a DSBH motif with significant similarity to that of FIH. Several JmjC containing histone demethylases have recently been identified that demethylate H3 Lys^{36} and/or H3 Lys^{9} [42]. Demethylation is thought to result from hydroxylation of the methyl group which forms a highly unstable intermediate that decomposes into formaldehyde plus a lysine residue (Figure 3). JmjC histone demethylases may therefore represent an even more direct link between oxygen availability and gene expression than the HIF-dependent regulation of HRE-containing promoters.

Conclusion

Studies into the regulation of EPO expression have led to the discovery of an oxygen-sensing system based on the activity of a set of 2-OG- and Fe(II)-dependent oxygenases and a transcriptional complex termed HIF that is central to gene regulation by hypoxia. The recognition that this enzyme family has many members, the discovery of other hydroxylated proteins and other substrates of the HIF hydroxylases, has raised the possibility that similar oxygen-sensitive signalling systems may operate on other processes. An important challenge will now be to define these pathways.

Summary

- *HIF is a major component of the hypoxia-induced response.*
- *HIF-α stability and activity are regulated by prolyl and asparaginyl hydroxylases respectively.*

- *The HIF hydroxylases belong to a large family of 2-OG/Fe(II)-dependent dioxygenases.*
- *Other substrates of the HIF hydroxylases have been identified, some of which may play a role in hypoxia-induced responses.*
- *Several other intracellular proteins are suggested to be, or have been shown to be, hydroxylated. In most cases the hydroxylases responsible are yet to be identified.*
- *2-OG/Fe(II)-dependent dioxygenases regulate diverse processes including DNA repair and histone demethylation, which may be hypoxia-responsive.*

References

1. Goda, N., Dozier, S.J. & Johnson, R.S. (2003) HIF-1 in cell cycle regulation, apoptosis, and tumor progression. *Antioxid. Redox Signal.* **5**, 467–473
2. Kaelin, W.G. (2005) Proline hydroxylation and gene expression. *Annu. Rev. Biochem.* **74**, 115–128
3. Pouyssegur, J., Dayan, F. & Mazure, N.M. Hypoxia signalling in cancer and approaches to enforce tumour regression. *Nature* **441**, 437–443
4. Pugh, C.W. & Ratcliffe, P.J. (2003) Regulation of angiogenesis by hypoxia: role of the HIF system. *Nat. Med.* **9**, 677–684
5. Schofield, C.J. & Ratcliffe, P.J. (2004) Oxygen sensing by HIF hydroxylases. *Nat. Rev. Mol. Cell Biol.* **5**, 343–354
6. Semenza, G.L. (1999) Regulation of mammalian O_2 homeostasis by hypoxia-inducible factor 1. *Annu. Rev. Cell. Dev. Biol.* **15**, 551–578
7. Semenza, G.L. (2003) Targeting HIF-1 for cancer therapy. *Nat. Rev. Cancer* **3**, 721–732
8. Wenger, R.H. (2002) Cellular adaptation to hypoxia: O_2-sensing protein hydroxylases, hypoxia-inducible transcription factors, and O_2-regulated gene expression. *FASEB J.* **16**, 1151–1162
9. Elvidge, G.P., Glenny, L., Appelhoff, R.J., Ratcliffe, P.J., Ragoussis, J. & Gleadle, J.M. (2006) Concordant regulation of gene expression by hypoxia and 2-oxoglutarate dependent dioxygenase inhibition; the role of HIF-1α, HIF-2α and other pathways. *J. Biol. Chem.* **281**, 15215–15226
10. McDonough, M.A., Li, V., Flashman, E., Chowdhury, R., Mohr, C., Lienard, B.M., Zondlo, J., Oldham, N.J., Clifton, I.J., Lewis, J. et al. (2006) Cellular oxygen sensing: crystal structure of hypoxia-inducible factor prolyl hydroxylase (PHD2). *Proc. Natl. Acad. Sci. U.S.A.* **103**, 9814–9819
11. Choi, K.O., Lee, T., Lee, N., Kim, J.H., Yang, E.G., Yoon, J.M., Kim, J.H., Lee, T.G. & Park, H. (2005) Inhibition of the catalytic activity of hypoxia-inducible factor-1α-prolyl-hydroxylase 2 by a MYND-type zinc finger. *Mol. Pharmacol.* **68**, 1803–1809
12. Schofield, C.J. & Ratcliffe, P.J. (2005) Signalling hypoxia by HIF hydroxylases. *Biochem. Biophys. Res. Commun.* **338**, 617–626
13. Berra, E., Benizri, E., Ginouves, A., Volmat, V., Roux, D. & Pouyssegur, J. (2003) HIF prolyl-hydroxylase 2 is the key oxygen sensor setting low steady-state levels of HIF-1α in normoxia. *EMBO J.* **22**, 4082–4090
14. Percy, M.J., Zhao, Q., Flores, A., Harrison, C., Lappin, T.R., Maxwell, P.H., McMullin, M.F. & Lee, F.S. (2006) A family with erythrocytosis establishes a role for prolyl hydroxylase domain protein 2 in oxygen homeostasis. *Proc. Natl. Acad. Sci. U.S.A.* **103**, 654–659
15. Takeda, K., Ho, V.C., Takeda, H., Duan, L.J., Nagy, A. & Fong, G.H. (2006) Placental but not heart defect is associated with elevated HIFα levels in mice lacking prolyl hydroxylase domain protein 2. *Mol. Cell. Biol.* **26**, 8336–8346
16. Appelhoff, R.J., Tian, Y.M., Raval, R.R., Turley, H., Harris, A.L., Pugh, C.W., Ratcliffe, P.J. & Gleadle, J.M. (2004) Differential function of the prolyl hydroxylases PHD1, PHD2 and PHD3 in the regulation of hypoxia-inducible factor. *J. Biol. Chem.* **279**, 38458–38465

17. Lieb, M.E., Menzies, K., Moschella, M.C., Ni, R. & Tubman, M.B. (2002) Mammalian EGLN genes have distinct patterns of mRNA expression and regulation. *Biochem. Cell Biol.* **80**, 421–426
18. Tian, Y.M., Mole, D.R., Ratcliffe, P.J. & Gleadle, J.M. (2006) Characterization of different isoforms of the HIF prolyl hydroxylase PHD1 generated by alternative initiation. *Biochem. J.* **397**, 179–186
19. Cervera, A.M., Apostolova, N., Luna-Crespo, F., Sanjuan-Pla, A., Garcia-Bou, R. & McCreath, K.J. (2006) An alternatively spliced transcript of the PHD3 gene retains prolyl hydroxylase activity. *Cancer Lett.* **233**, 131–138
20. Hirsila, M., Koivunen, P., Gunzler, V., Kivirikko, K.I. & Myllyharju, J. (2003) Characterization of the human prolyl 4-hydroxylases that modify the hypoxia-inducible factor HIF. *J. Biol. Chem.* **278**, 30772–30780
21. Nakayama, K., Frew, I.J., Hagensen, M., Skals, M., Habelhah, H., Bhoumik, A., Kadoya, T., Erdjument-Bromage, H., Tempst, P., Frappell, P.B. et al. (2004) Siah2 regulates stability of prolyl-hydroxylases, controls HIF1α abundance, and modulates physiological responses to hypoxia. *Cell* **117**, 941–952
22. Masson, N., Appelhoff, R.J., Tuckerman, J.R., Tian, Y.M., Demol, H., Puype, M., Vandekerckhove, J., Ratcliffe, P.J. & Pugh, C.W. (2004) The HIF prolyl hydroxylase PHD3 is a potential substrate of the TRiC chaperonin. *FEBS Lett.* **570**, 166–170
23. Baek, J.H., Mahon, P.C., Oh, J., Kelly, B., Krishnamachary, B., Pearson, M., Chan, D.A., Giacca, A.J. & Semenza, G.L. (2005) OS-9 interacts with hypoxia-inducible factor 1α and prolyl hydroxylases to promote oxygen-dependent degradation of HIF-1α. *Mol. Cell* **17**, 503–512
24. Hopfer, U., Hopfer, H., Jablonski, K., Stahl, R.A. & Wolf, G. (2006) The novel WD-repeat protein Morg1 acts as a molecular scaffold for hypoxia-inducible factor prolyl hydroxylase 3 (PHD3). *J. Biol. Chem.* **281**, 8645–8655
25. Hewitson, K.S., Granatino, N., Welford, R.W., McDonough, M.A. & Schofield, C.J. (2005) Oxidation by 2-oxoglutarate oxygenases: non-haem iron systems in catalysis and signalling. *Philos. Transact. A Math. Phys. Eng. Sci.* **363**, 807–828; discussion 1035–1040
26. King, A., Selak, M.A. & Gottlieb, E. (2006) Succinate dehydrogenase and fumarate hydratase: linking mitochondrial dysfunction and cancer. *Oncogene* **25**, 4675–4682
27. Gerald, D., Berra, E., Frapart, Y.M., Chan, D.A., Giacca, A.J., Mansuy, D., Pouyssegur, J., Yaniv, M. & Mechta-Grigoriou, F. (2004) JunD reduces tumor angiogenesis by protecting cells from oxidative stress. *Cell* **118**, 781–794
28. Kaelin, Jr, W.G. (2005) ROS: really involved in oxygen sensing. *Cell Metab.* **1**, 357–358
29. Stolze, I.P., Tian, Y.M., Appelhoff, R.J., Turley, H., Wykoff, C.C., Gleadle, J.M. & Ratcliffe, P.J. (2004) Genetic analysis of the role of the asparaginyl hydroxylase FIH in regulating HIF transcriptional target genes. *J. Biol. Chem.* **279**, 42719–42725
30. Bracken, C.P., Fedele, A.O., Linke, S., Balrak, W., Lisy, L., Whitelaw, M.L. & Peet, D.J. (2006) Cell-specific regulation of hypoxia-inducible factor (HIF)-1α and HIF-2α stabilization and transactivation in a graded oxygen environment. *J. Biol. Chem.* **281**, 22575–22585
31. Dayan, F., Brahimi-Horn, M.C., Pouyssegur, J. & Mazure, N.M. (2006) The oxygen sensor factor-inhibiting hypoxia-inducible factor-1 controls expression of distinct genes through the bifunctional transcriptional character of hypoxia-inducible factor-1α. *Cancer Res.* **66**, 3688–3698
32. Ehrismann, D., Flashman, E., Genn, D.N., Mathioudakis, N., Hewitson, K.S., Ratcliffe, P.J. & Schofield, C.J. (2007) Studies on the activity of the hypoxia-inducible factor hydroxylases using an oxygen consumption assay. *Biochem. J.* **401**, 227–234
33. Koivunen, P., Hirsila, M., Kivirikko, K.I. & Myllyharju, J. (2006) The length of peptide substrates has a marked effect on hydroxylation by the hypoxia-inducible factor prolyl 4-hydroxylases. *J. Biol. Chem.* **281**, 28712–28720
34. Seth, P., Krop, I., Porter, D. & Polyak, K. (2002) Novel estrogen and tamoxifen induced genes identified by SAGE (Serial Analysis of Gene Expression). *Oncogene* **21**, 836–843
35. Frei, C. and Edgar, B.A. (2004) Drosophila cyclin D/Cdk4 requires HIF-1 prolyl hydroxylase to drive cell growth. *Develop. Cell* **6**, 241–251
36. Lee, S., Nakamura, E., Yang, H., Wei, W., Linggi, M.S., Sajan, M.P., Farese, R.V., Freeman, R.S., Carter, B.D., Kaelin, Jr, W.G. & Schlisio, S. (2005) Neuronal apoptosis linked to EglN3 prolyl

hydroxylase and familial pheochromocytoma genes: developmental culling and cancer. *Cancer Cell* **8**, 155–167

37. Straub, J.A., Lipscomb, E.A., Yoshida, E.S. & Freeman R.S. (2003) Induction of SM-20 in PC12 cells leads to increased cytochrome *c* levels, accumulation of cytochrome *c* in the cytosol, and caspase-dependent cell death. *J. Neurochem.* **85**, 318–328
38. Xie, L., Johnson, R.S. & Freeman, R.S. (2005) Inhibition of NGF deprivation-induced death by low oxygen involves suppression of BIMEL and activation of HIF-1. *J. Cell Biol.* **168**, 911–920
39. Cummins, E.P., Berra, E., Comerford, K.M., Ginouves, A., Fitzgerald, K.T., Seeballuck, F., Godson, C., Nielsen, J.E., Moynagh, P., Pouyssegur, J. & Taylor, C.T. (2006) Prolyl hydroxylase-1 negatively regulates IκB kinase-β, giving insight into hypoxia-induced NFκB activity. *Proc. Natl. Acad. Sci. U.S.A.* **103**, 18154–18159
40. Cockman, M.E., Lancaster, D.E., Stolze, I.P., Hewitson, K.S., McDonough, M.A., Coleman, M.L., Coles, C.H., Yu, X., Hay, R.T., Ley, S.C et al. (2006) Posttranslational hydroxylation of ankyrin repeats in IκB proteins by the hypoxia-inducible factor (HIF) asparaginyl hydroxylase, factor inhibiting HIF (FIH). *Proc. Natl. Acad. Sci. U.S.A.* **103**, 14767–14772
41. Mishina, Y. & He, C. (2006) Oxidative dealkylation DNA repair mediated by the mononuclear non-heme iron AlkB proteins. *J. Inorg. Biochem.* **100**, 670–678
42. Klose, R.J., Kallin, E.M. & Zhang, Y. (2006) JmjC-domain-containing proteins and histone demethylation. *Nat. Rev. Genet.* **7**, 715–727

2

Mitochondrial oxygen sensing: regulation of hypoxia-inducible factor by mitochondrial generated reactive oxygen species

Eric L. Bell* and Navdeep S. Chandel*†[1]

**Department of Medicine, Northwestern University Medical School, Chicago, IL 60611, U.S.A. and †Department of Cell and Molecular Biology, Northwestern University Medical School, Chicago, IL 60611, U.S.A.*

Abstract

Decreased oxygen availability (hypoxia) promotes physiological processes such as energy metabolism, angiogenesis, cell proliferation and cell viability through the transcription factor HIF (hypoxia-inducible factor). Activation of HIF can also promote pathophysiological processes such as cancer and pulmonary hypertension. The mechanism(s) by which hypoxia activates HIF are the subject of intensive research. In this chapter we outline the model in which mitochondria regulate the stability of HIF through the increased production of ROS (reactive oxygen species) during hypoxia.

Introduction

Oxygen is necessary for survival of most eukaryotes. Just as carbon sources are used by glycolysis to generate energy, cells utilize oxygen to generate

[1]*To whom correspondence should be addressed (email nav@northwestern.edu).*

the majority of ATP required for normal cellular processes through oxidative phosphorylation [1]. In the complete absence of oxygen (anoxia, 0–0.3% oxygen), most organisms and cells cannot survive for long periods of time [2,3]. However, in conditions of limited oxygen (hypoxia, 0.3–5% oxygen), cells activate an adaptive response that is mediated by the transcription factor HIF (hypoxia-inducible factor) [4]. This response facilitates survival through the up-regulation of a number of genes involved in various physiological processes such as glycolysis and angiogenesis that promote survival in hypoxic conditions by increasing the capacity for generating ATP and inhibiting the depletion of oxygen. This adaptation response is not due to compromised bioenergetics during hypoxia. The activity of the terminal complex in oxidative phosphorylation, cytochrome *c* oxidase, is not significantly decreased, and therefore cellular ATP levels do not change in hypoxic conditions. The K_m of cytochrome *c* oxidase is less than 0.3% oxygen. In contrast with hypoxia, cytochrome *c* oxidase activity is substantially decreased during anoxia resulting in a decrease in the generation of ATP [5].

The transcriptional activation of HIF is a stress response developed through evolution to allow for cells to avoid a bioenergetic crisis in low oxygen levels, making HIF a master regulator of oxygen homoeostasis. An example of organismal maintenance of oxygen homoeostasis in higher organisms is their ability to increase their capacity for systemic oxygen delivery. This requires erythropoiesis, the process of increased generation of erythrocytes. Erythropoiesis is increased under conditions of decreased oxygen availability through an increase in expression of the protein erythropoietin, a HIF target gene [6]. In fact early examination of the regulation of erythropoietin in response to oxygen led to the discovery of the transcription factor HIF.

Oxygen homoeostasis is also critical for proper development, making HIF-mediated transcription necessary for normal organismal development. The fact that HIF knockout mice are embryonic lethal at embryonic day 9.5 highlights the importance of HIF-mediated transcription in development [7,8]. The lethality displayed by HIF knockout mice is due to impaired vascular development and abnormal placental development [9]. An important HIF target gene is VEGF (vascular endothelial growth factor), which promotes neovascularization and is required for development [10,11]. Unfortunately, genes such as VEGF and other HIF target genes that are required for normal maintenance of oxygen homoeostasis also promote pathophysiological processes such as pulmonary hypertension and cancer. Studies have indicated that preventing HIF activation can suppress tumorigenesis or hypoxia-induced pulmonary hypertension [12–15]. Therefore defining how a cell senses decreased levels of oxygen to regulate the activity of HIF has broad implications for normal physiology as well as for numerous diseases associated with hypoxia. Genetic studies from our laboratory and others support a model in which hypoxia induces mitochondria to increase the generation of ROS (reactive oxygen species) that are both necessary and sufficient to activate the HIF-mediated transcriptional programme for adaptation to hypoxia [16–18]. This chapter focuses on the role of mitochondrial ROS as a key component of oxygen sensing.

Oxygen regulation of HIF

HIF is a heterodimer of two basic helix-loop-helix/PAS (Per-Arnt-Sim) proteins, HIF-α and the ARNT (aryl hydrocarbon receptor nuclear translocator) or HIF-1β [19]. To date three HIF-α isoforms have been described, HIF-1α, HIF-2α and HIF-3α. HIF-1α and HIF-2α are transcriptionally active [20–21]. Their target genes have been described to be both overlapping and distinct. The function of HIF-3α is less well-defined, however it is thought to be an antagonist for HIF-1α- and HIF-2α-mediated transcription [22]. Both HIF-α and ARNT protein subunits are expressed ubiquitously, but the stability of each protein is differentially regulated by oxygen levels. Initial studies demonstrated that HIF-α protein levels are labile at normal oxygen conditions, whereas HIF-β protein levels are constitutively stable [23]. Recent studies indicate that oxygen levels regulate the hydroxylation of two proline residues, 402 and 564, within the ODDD (oxygen-dependent degradation domain) of HIF-α [24]. This hydroxylation reaction is catalysed by a family of PHDs (proline hydroxylation enzymes) [25,26]. The PHDs require Fe^{2+}, oxygen and 2-oxoglutarate to catalyse the hydroxylation reaction. Hydroxylated proline serves as a binding site for the VHL (von Hippel-Lindau) protein, the substrate recognition component of the VBC-CUL-2 E3 ubiquitin ligase complex [27–29]. Once bound, pVHL tags HIF-α with ubiquitin thereby targeting it for proteasomal degradation [30]. HIF-α also contains two transactivation domains. The NAD (N-terminal transactivation domain) is located within the ODDD. The CAD (C-terminus transactivation domain) contains an asparagine residue that is hydroxylated by FIH (factor inhibiting HIF) [31]. This reaction takes place under normoxic conditions (21% oxygen) and it inhibits the transactivation potential of HIF-α [32]. This is due to the inability of HIF-α to interact with the co-activator CBP/p300 as a result of the hydroxy group on Asn^{803} [33].

When oxygen levels decrease below 5%, HIF-α protein is stabilized due to a lack of proline hydroxylation (Figure 1) in the absence of proline hydroxylation, VHL protein cannot bind HIF-α to initiate ubiquitin-proteasomal degradation. When stabilized, HIF-α translocates to the nucleus and dimerizes with HIF-β. Once in the nucleus, the HIF dimer binds to HREs (HIF response elements) located throughout the genome [34]. The absence of a hydroxy group on Asn^{803} allows HIF to associate with the co-activator CBP/p300 to facilitate the transcription of various target genes. Therefore regulation of HIF activity is tightly controlled by the hydroxylation of various amino acids. Understanding how the hydroxylation enzymes are regulated by hypoxia would provide the next step in elucidating the hypoxic response.

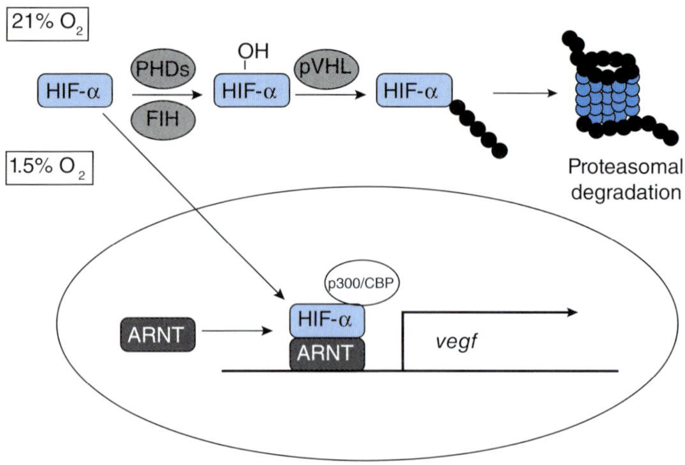

Figure 1. HIF-1 is composed of two subunits, oxygen sensitive HIF-1α and HIF-1β (ARNT)

Under normoxia HIF-1α is hydroxylated at two different proline residues and an asparagine residue. The hydroxylation of proline residues serves as a recognition motif for VHL protein (pVHL). The binding of VHL protein targets the HIF-1α protein for ubiquitin-mediated degradation. HIF-1α contains two transactivation domains referred to as NAD (N-TAD; amino acids 531–575) and CAD (C-TAD; amino acids 786–826). The asparagine residue resides in the CAD. The hydroxylation of asparagine prevents the binding of transcriptional co-activators such as p300/CBP. Under hypoxia, the hydroxylation of the proline and asparagine residues is diminished, which allows for the protein to be stabilized and bind to HIF-1β as well as p300/CBP to allow HIF-1-dependent gene transcription.

Mitochondrial electron transport and HIF

The mitochondria are responsible for the majority of the oxygen consumed within the cell and evolutionarily poised to be a likely target in the hunt for the cellular oxygen sensor. Mitochondria contain their own DNA (termed mtDNA) which encodes 13 genes that are essential for the assembly of a functional electron transport chain. These genes encode subunits for complexes I, III, IV and V. The major consumer of oxygen is complex IV. However, initial studies indicated that pharmacological inhibition of complex IV by cyanide did not activate or repress HIF [24]. Cells cultured with sub-lethal levels of ethidium bromide, which inhibits the transcription and replication of mtDNA resulting in the loss of a functional electron transport chain, were used to genetically address whether mitochondria are involved in the HIF response [35]. These cells devoid of mtDNA, ρ^0 cells, are unable to activate HIF-α in hypoxic conditions [36,37]. However, these cells are still able to stabilize HIF-α in anoxic conditions, indicating that there is a fundamental difference in the mechanism required to stabilize HIF-α protein in hypoxic compared with anoxic conditions [3]. These data indicate that the ability of mitochondria to perform electron transport is necessary for hypoxic stabilization of HIF-α but not anoxic stabilization. The mechanism utilized in anoxic conditions is most likely the direct inhibition of the PHDs due to a lack of oxygen. Since the PHDs require oxygen as a co-substrate, they cannot hydroxylate HIF-α in anoxic conditions to initiate degradation, thereby stabilizing HIF-α protein.

After initial studies demonstrating that ρ^0 cells fail to activate HIF during hypoxia, there were reports with conflicting data indicating that hypoxic activation of HIF is intact in ρ^0 cells [38–40]. However these data were most like due to the use of oxygen levels closer to anoxic conditions than hypoxic conditions. Alternatively, these data could be due to a difference in ρ^0 cells generated by ethidium bromide. Recently three independent reports utilized rigorous genetic methods to manipulate mitochondrial electron transport system to determine the requirement of a functional mitochondrial electron transport in hypoxic stabilization of HIF-α protein. The first study used mouse genetics to knockout cytochrome *c* [18]. In the electron transport chain, cytochrome *c* accepts electrons from complex III and transfers them to complex IV to be used to reduce molecular oxygen to water. Murine embryonic cells lacking cytochrome *c* failed to stabilize HIF-α during hypoxic conditions. The other two studies demonstrated that inhibiting complex III function by knocking down the Rieske Fe-S protein inhibits the ability of multiple cell lines to stabilize HIF-α in hypoxic conditions [16,17]. However, cells that displayed knockdown of the Rieske Fe-S protein were able to stabilize HIF-α in anoxic conditions, further supporting the notion that different mechanisms allow for the stabilization of HIF-α protein in hypoxic and anoxic conditions.

These data provide conclusive evidence that a functional electron transport chain is required for the hypoxic stabilization of HIF-α protein. However, the manner by which a functional electron transport chain activates HIF during hypoxia remains controversial. The loss of a functional electron transport chain prevents ROS generation as well as oxygen consumption and both have been proposed as potential mechanisms by which the mitochondrial electron transport chain activates HIF during hypoxia. Both of these mechanisms are discussed below.

Mitochondria regulate HIF dependent on respiration

Molecular oxygen is used as the terminal electron acceptor in the mitochondrial electron transport chain when cytochrome *c* oxidase converts oxygen into water [1]. This property of mitochondria combined with the requirement of the PHDs for molecular oxygen as a co-substrate led some groups to propose a model in which mitochondrial oxygen consumption is the regulator of HIF activation via the PHDs. These groups propose that during conditions of limited oxygen, mitochondria create an oxygen gradient within the cells as a result of their ability to consume oxygen [41,42]. This gradient would effectively sequester molecular oxygen away from the cytosolic PHDs, thus inhibiting their ability to hydroxylate HIF-α. In contrast, our studies have provided genetic experiments that demonstrate that HIF-α protein stability is independent of respiration in hypoxic conditions. Cells lacking a functional complex IV due to a mutation in the *surf* gene which encodes for a protein required for the proper assembly of complex IV, retain the ability to stabilize HIF-α protein in hypoxic conditions [16].

These cells that lack complex IV are respiratory deficient and cannot generate a cytosolic oxygen gradient to sequester molecular oxygen from the PHDs. Thus presently there is equivocal data on the role of mitochondrial respiration in the regulation of HIF during hypoxia.

Mitochondria regulate HIF dependent on ROS

Initial studies in exploring the mechanism by which mitochondria regulate the stability of HIF-α demonstrated that ROS levels paradoxically increase in hypoxic conditions [36]. The increase in ROS production during hypoxia is required and sufficient to stabilize HIF-α protein [37]. Cells deficient in cytochrome *c* or Rieske Fe-S protein which are unable to stabilize HIF during hypoxia also do not display an increase in ROS generation during hypoxia. Incubating cells with pharmacological antioxidants such as ebselen and MitoQ attenuates HIF activation in hypoxic conditions [17,43]. Expression of protein antioxidants such as GPX (glutathione peroxidase) and catalase also attenuates HIF-α protein stabilization and activation, but expression of SOD (superoxide dismutase) has no effect [16]. SOD converts superoxide into H_2O_2, while GPX and catalase converts H_2O_2 into water. The specificities of these antioxidants for different forms of ROS tends toward the conclusion that H_2O_2 is the ROS moiety required for stabilization of HIF-α protein. In fact HIF-α protein is stabilized when cells are pulsed with 25μ t-butyl H_2O_2, a more stable form of H_2O_2, in normal oxygen conditions, indicating that H_2O_2 is also sufficient to activate HIF-mediated transcription [37]. ROS are a normal by-product of electron transport within the mitochondria, therefore the ability of cells to increase the generation of ROS in hypoxic conditions provides the link between mitochondrial electron transport and HIF activation.

A major controversy with the ROS model with respect to hypoxic activation of HIF is whether hypoxia increases ROS. There have been multiple studies that have demonstrated both an increase and a decrease in ROS production during hypoxia [36,38,44–48]. All of these studies utilized dyes which have limitations in their measurements of intracellular oxidative stress [49]. To resolve whether hypoxia increases or decreases ROS, Guzy et al. [17] have utilized a sensitive FRET (fluorescence resonance energy transfer) probe to assess redox status in the cytosol during hypoxia. This probe consists of a fusion protein containing two fluorescent peptides (CFP and YFP) linked by a redox-sensitive hinge that contains cysteine thiols that become cross-linked by oxidant stress. When expressed in cells the redox-sensitive FRET probe responds to hypoxia by producing a dose-dependent increase in FRET ratio. FRET can be difficult to use, thus the recent development of redox-sensitive GFP (green fluorescent protein) probes will help resolve whether hypoxia increases or decreases ROS. Furthermore, Pan et al. [50] have shown that MitoQ can inhibit respiration thus making it difficult to discern whether MitoQ works as a respiratory inhibitor or an antioxidant. Future genetic studies need to uncouple ROS generation from oxygen consumption to address which function of the mitochondrial electron transport chain is the major regulator of HIF during hypoxia.

Mitochondria and PHDs

The most proximal regulators of HIF activation are the PHDs. How mitochondrial-generated ROS would cross-talk to the PHDs is not presently known. The addition of exogenous H_2O_2 to cells under normal oxygen conditions stabilizes the HIF-α protein, implying that ROS inhibit the hydroxylation of the HIF-α protein. Presently there are two mechanisms by which ROS could prevent hydroxylation of HIF-1α protein. One possibility is that low levels of oxygen decrease PHD activity while the ROS decrease the availability of the co-factor Fe^{2+}. Indeed Gerald et al. [51] found that loss of JunD caused ROS to accumulate which decreased the availability of Fe^{2+} and reduced the activity of HIF-PHDs [51]. A second possibility is that ROS activate signalling pathways that catalytically make PHDs inactive. Multiple signalling pathways have been implicated in hypoxic stabilization of HIF [52–55]. Finally, it could be that the low oxygen levels decrease both PHD activity and increase the ROS produced during hypoxia. These two events coupled together decrease PHD activity to levels that prevent hydroxylation of HIF-α protein during hypoxia (Figure 2). Further studies need to address how the mitochondria and PHDs crosstalk.

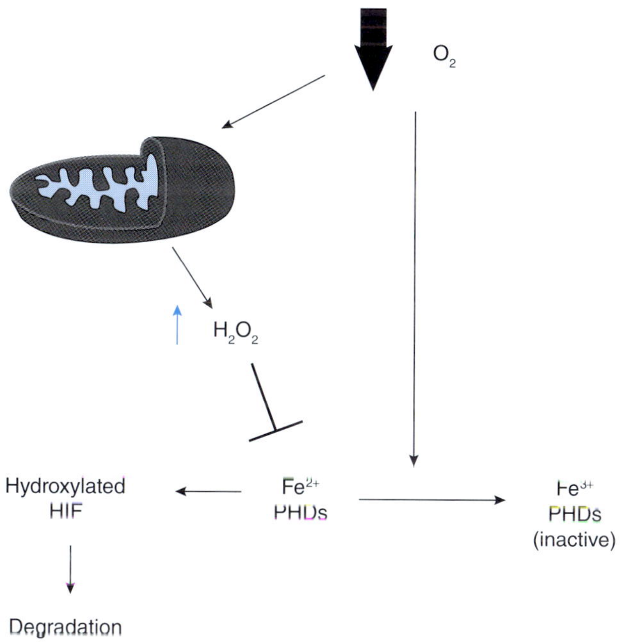

Figure 2. Hypoxia directly decreases the activity of PHDs which is coupled with an increase in ROS generation; this depresses PHD activity therefore inhibiting hydroxylation of the HIF-α protein

Concluding remarks

HIF is the master regulator of oxygen homoeostasis, with key roles in various physiological processes especially during development. Aberrant HIF activation leads to pathophysiological processes, most notably cancer, thereby demonstrating the importance of understanding how HIF is regulated by oxygen. Existing literature clearly demonstrates that mitochondria regulate HIF activation during hypoxia. However, it is not clear whether it is the ability of the mitochondria to generate ROS or simply consume oxygen that is the major requirement in regulation of HIF. Furthermore, additional data are required to understand how to couple mitochondria to regulation of PHD activity during hypoxia. Understanding this pathway will provide insight into how hypoxic signalling can be modified in disease to promote physiological processes as well as how to treat disease by hindering the pathophysiological processes governed by HIF.

Summary

- *Decreased oxygen availability (hypoxia) promotes physiological processes such as energy metabolism, angiogenesis, cell proliferation and cell viability through the transcription factor HIF.*
- *Activation of HIF can also promote pathophysiological processes like cancer and pulmonary hypertension.*
- *HIF-1 is composed of two subunits, oxygen sensitive HIF-1α and HIF-β. HIF-1α is hydroxylated at two different proline residues and an asparagine residue under normoxia. The hydroxylation of proline residues serves as a recognition motif for VHL protein. The binding of VHL protein targets the HIF-1α protein for ubiquitin-mediated degradation under hypoxia, the hydroxylation of proline and asparagine is diminished, which allows for the protein to be stabilized and bind to HIF-1β as well as p300/CBP to allow HIF-1-dependent gene transcription.*
- *Multiple studies have demonstrated that the mitochondrial electron transport chain regulates the activation of HIF during hypoxia. Presently, there are two models to explain the mitochondrial regulation of HIF.*
- *The first model states that mitochondria create an oxygen gradient within the cells as a result of their ability to consume oxygen. This gradient effectively sequesters molecular oxygen away from the cytosolic PHDs, thus inhibiting their ability to hydroxylate HIF-α.*
- *The second model postulates that mitochondria generate ROS which are required and sufficient to activate HIF during hypoxia.*
- *Future genetic studies need to uncouple ROS generation from oxygen consumption to address which function of the mitochondrial electron transport chain is the major regulator of HIF during hypoxia.*

References

1. Saraste, M. (1999) Oxidative phosphorylation at the fin de siecle. *Science* **283**, 1488–1493
2. Graeber, T.G., Osmanian, C., Jacks, T., Housman, D.E., Koch, C.J., Lowe, S.W. & Giaccia, A.J. (1996) Hypoxia-mediated selection of cells with diminished apoptotic potential in solid tumours. *Nature* **379**, 88–91
3. Schroedl, C., McClintock, D.S., Budinger, G.R.S. & Chandel, N.S. (2002) Hypoxic but not anoxic stabilization of HIF-1α requires mitochondrial reactive oxygen species. *Am. J. Physiol. Lung Cell Mol. Physiol.* **283**, L922–L931
4. Semenza, G.L. (1998) Hypoxia-inducible factor 1 and the molecular physiology of oxygen homeostasis. *J. Lab. Clin. Med.* **131**, 207–214
5. Chandel, N.S., Budinger, G.R. & Schumacker, P.T. (1996) Molecular oxygen modulates cytochrome c oxidase function. *J. Biol. Chem.* **271**, 18672–18677
6. Semenza, G.L. (1994) Regulation of erythropoietin production. New insights into molecular mechanisms of oxygen homeostasis. *Hematol. Oncol. Clin. North Am.* **8**, 863–884
7. Ryan, H.E., Lo, J. & Johnson, R.S. (1998) HIF-1α is required for solid tumor formation and embryonic vascularization. *EMBO J.* **17**, 3005–3015
8. Maltepe, E., Schmidt, J.V., Baunoch, D., Bradfield, C.A. & Simon, M.C. (1997) Abnormal angiogenesis and responses to glucose and oxygen deprivation in mice lacking the protein ARNT. *Nature* **386**, 403–407
9. Adelman, D.M., Maltepe, E. & Simon, M.C. (2000) Placental cell fates are regulated *in vivo* by HIF-mediated hypoxia responses. *Genes Dev.* **14**, 3191–3203
10. Forsythe, J.A., Jiang, B.H., Iyer, N.V., Agani, F., Leung, S.W., Koos, R.D. & Semenza, G.L. (1996) Activation of vascular endothelial growth factor gene transcription by hypoxia-inducible factor 1. *Mol. Cell. Biol.* **16**, 4604–4613
11. Breier, G., Albrecht, U., Sterrer, S. & Risau, W. (1992) Expression of vascular endothelial growth factor during embryonic angiogenesis and endothelial cell differentiation. *Development* **114**, 521–532
12. Yu, A.Y., Shimoda, L.A., Iyer, N.V., Huso, D.L., Sun, X., McWilliams, R., Beaty, T., Sham, J.S., Wiener, C.M., Sylvester, J.T. & Semenza, G.L. (1999) Impaired physiological responses to chronic hypoxia in mice partially deficient for hypoxia-inducible factor 1α. *J. Clin. Invest.* **103**, 691–696
13. Maranchie, J.K., Vasselli, J.R., Riss, J., Bonifacino, J.S., Linehan, W.M. & Klausner, R.D. (2002) The contribution of VHL substrate binding and HIF1-α to the phenotype of VHL loss in renal cell carcinoma. *Cancer Cell* **1**, 247–255
14. Kondo, K., Klco, J., Nakamura, E., Lechpammer, M. & Kaelin, Jr, W.G. (2002) Inhibition of HIF is necessary for tumor suppression by the von Hippel-Lindau protein. *Cancer Cell* **1**, 237–246
15. Kung, A.L., Wang, S., Klco, J.M. & Livingston, D.M. (2000) Suppression of tumor growth through disruption of hypoxia-inducible transcription. *Nat. Med.* **6**, 1335–1340
16. Brunelle, J.K., Bell, E.L., Quesada, N.M., Vercauteren, K., Tiranti, V., Zeviani, M., Scarpulla, R.C. & Chandel, N.S. (2005) Oxygen sensing requires mitochondrial ROS but not oxidative phosphorylation. *Cell Metab.* **1**, 409–414
17. Guzy, R.D., Hoyos, B., Robin, E., Chen, H., Liu, L., Mansfield, K.D., Simon, M.C., Hammerling, U. & Schumacker, P.T. (2005) Mitochondrial complex III is required for hypoxia-induced ROS production and cellular oxygen sensing. *Cell Metab.* **1**, 401–408
18. Mansfield, K.D., Guzy, R.D., Pan Y., Young, R.M., Cash, T.P., Schumacker, P.T. & Simon, M.C. (2005) Mitochondrial dysfunction resulting from loss of cytochrome c impairs cellular oxygen sensing and hypoxic HIF-α activation. *Cell Metab.* **1**, 393–399
19. Wang, G.L., Jiang, B.H., Rue, E.A. & Semenza, G.L. (1995) Hypoxia-inducible factor 1 is a basic-helix-loop-helix-PAS heterodimer regulated by cellular O_2 tension. *Proc. Natl. Acad. Sci. U.S.A.* **92**, 5510–5514
20. O'Rourke, J.F., Tian, Y.M., Ratcliffe, P.J. & Pugh, C.W. (1999) Oxygen-regulated and transactivating domains in endothelial PAS protein 1: comparison with hypoxia-inducible factor-1α. *J. Biol. Chem.* **274**, 2060–2071

21. Wiesener, M.S., Turley, H., Allen, W.E., Willam, C., Eckardt, K.U., Talks, K.L., Wood, S.M., Gatter, K.C., Harris, A.L., Pugh, C.W., Ratcliffe, P.J. & Maxwell, P.H. (1998) Induction of endothelial PAS domain protein-1 by hypoxia: characterization and comparison with hypoxia-inducible factor-1α. *Blood* **92**, 2260–2268
22. Makino, Y., Cao, R., Svensson, K., Bertilsson, G., Asman, M., Tanaka, H., Cao, Y., Berkenstam, A. & Poellinger, L. (2001) Inhibitory PAS domain protein is a negative regulator of hypoxia-inducible gene expression. *Nature* **414**, 550–554
23. Jiang, B.H., Semenza, G.L., Bauer, C., Marti, H.H. (1996) Hypoxia-inducible factor 1 levels vary exponentially over a physiologically relevant range of O_2 tension. *Am. J. Physiol.* **271**, C1172–C1180
24. Masson, N., Willam, C., Maxwell, P.H., Pugh, C.W. & Ratcliffe, P.J. (2001) Independent function of two destruction domains in hypoxia-inducible factor-α chains activated by prolyl hydroxylation. *EMBO J.* **20**, 5197–5206
25. Bruick, R.K. & S.L. McKnight (2001) A conserved family of prolyl-4-hydroxylases that modify HIF. *Science* **294**, 1337–1340
26. Epstein, A.C., Gleadle, J.M., McNeill, L.A., Hewitson, K.S., O'Rourke, J., Mole, D.R., Mukherji, M., Metzen, E., Wilson, M.I., Dhanda, A. et al. (2001) C. elegans EGL-9 and mammalian homologs define a family of dioxygenases that regulate HIF by prolyl hydroxylation. *Cell* **107**, 43–54
27. Iwai, K., Yamanaka, K., Kamura, T., Minato, N., Conaway, R.C., Conaway, J.W., Klausner, R.D. & Pause A. (1999) Identification of the von Hippel-Lindau tumor-suppressor protein as part of an active E3 ubiquitin ligase complex. *Proc. Natl. Acad. Sci. U.S.A.* **96**, 12436–12441
28. Jaakkola, P., Mole, D.R., Tian, Y.M., Wilson, M.I., Gielbert, J., Gaskell, S.J., Kriegsheim, A.v., Hebestreit, H.F., Mukherji, M., Schofield, C.J. et al. (2001) Targeting of HIF-alpha to the von Hippel-Lindau ubiquitylation complex by O_2-regulated prolyl hydroxylation. *Science* **292**, 468–472
29. Ivan, M., Kondo, K., Yang, H., Kim, W., Valiando, J., Ohh, M., Salic, A., Asara, J.M., Lane, W.S. & Kaelin, Jr, W.G. (2001) HIFα targeted for VHL-mediated destruction by proline hydroxylation: implications for O_2 sensing. *Science* **292**, 464–468
30. Maxwell, P.H., Wiesener, M.S., Chang, G.W., Clifford, S.C., Vaux, E.C., Cockman, M.E., Wykoff, C.C., Pugh, C.W., Maher, E.R. & Ratcliffe, P.J. (1999) The tumour suppressor protein VHL targets hypoxia-inducible factors for oxygen-dependent proteolysis. *Nature* **399**, 271–275
31. Mahon, P.C., Hirota K. & Semenza G.L. (2001) FIH-1: a novel protein that interacts with HIF-1α and VHL to mediate repression of HIF-1 transcriptional activity. *Genes Dev.* **15**, 2675–2686
32. Lando, D., Peet, D.J., Whelan, D.A., Gorman, J.J. & Whitelaw, M.L. (2002) FIH-1 is an asparaginyl hydroxylase enzyme that regulates the transcriptional activity of hypoxia-inducible factor. *Genes Dev.* **16**, 1466–1471
33. Freedman, S.J., Sun, Z.Y., Poy, F., Kung, A.L., Livingston, D.M., Wagner, G. & Eck, M.J. (2002) Structural basis for recruitment of CBP/p300 by hypoxia-inducible factor-1α. *Proc. Natl. Acad. Sci. U.S.A.* **99**, 5367–5372
34. Jiang, B.H., Zheng, J.Z., Leung, S.W., Roe, R. & Semenza, G.L. (1997) Transactivation and inhibitory domains of hypoxia-inducible factor 1α modulation of transcriptional activity by oxygen tension. *J. Biol. Chem.* **272**, 19253–19260
35. King, M.P. & Attardi G. (1988) Injection of mitochondria into human cells leads to a rapid replacement of the endogenous mitochondrial DNA. *Cell* **52**, 811–819
36. Chandel, N.S., Maltepe, E., Goldwasser, E., Mathieu, C.E., Simon, M.C. & Schumacker, P.T. (1998) Mitochondrial reactive oxygen species trigger hypoxia-induced transcription. *Proc. Natl. Acad. Sci. U.S.A.* **95**, 11715–11720
37. Chandel, N.S., McClintock, D.S., Feliciano, S.E., Wood, T.M., Melendez, J.A., Rodriguez, A.M. & Schumacker, P.T. (2000) Reactive oxygen species generated at mitochondrial complex III stabilize hypoxia-inducible factor-1α during hypoxia: a mechanism of O_2 sensing. *J. Biol. Chem.* **275**, 25130–25138
38. Vaux, E.C., Metzen, E., Yeates, K.M. & Ratcliffe, P.J. (2001) Regulation of hypoxia-inducible factor is preserved in the absence of a functioning mitochondrial respiratory chain. *Blood* **98**, 296–302

39. Enomoto, N., Koshikawa, N., Gassmann, M., Hayashi, J. & Takenaga, K. (2002) Hypoxic induction of hypoxia-inducible factor-1α and oxygen-regulated gene expression in mitochondrial DNA-depleted HeLa cells. *Biochem. Biophys. Res. Commun.* **297**, 346–352
40. Srinivas, V., Zhu, X., Salceda, S., Nakamura, R. & Caro, J. (2001) Oxygen sensing and HIF-1 activation does not require an active mitochondrial respiratory chain electron-transfer pathway. *J. Biol. Chem.* **276**, 21995–21998
41. Hagen, T., Taylor, C.T., Lam, F. & Moncada, S. (2003) Redistribution of intracellular oxygen in hypoxia by nitric oxide: effect on HIF1α. *Science* **302**, 1975–1978
42. Doege, K., Heine, S., Jensen, I., Jelkmann, W. & Metzen E. (2005) Inhibition of mitochondrial respiration elevates oxygen concentration but leaves regulation of hypoxia-inducible factor (HIF) intact. *Blood* **106**, 2311–2317
43. Sanjuan-Pla, A., Cervera, A.M., Apostolova, N., Garcia-Bou, R., Victor, V.M., Murphy, M.P. & McCreath, K.J. (2005) A targeted antioxidant reveals the importance of mitochondrial reactive oxygen species in the hypoxic signaling of HIF-1α. *FEBS Lett.* **579**, 2669–2674
44. Duranteau, J., Chandel, N.S., Kulisz, A., Shao, Z. & Schumacker, P.T. (1998) Intracellular signaling by reactive oxygen species during hypoxia in cardiomyocytes. *J. Biol. Chem.* **273**, 11619–11624
45. Killilea, D.W., Hester, R., Balczon, R., Babal, P. & Gillespie, M.N. (2000) Free radical production in hypoxic pulmonary artery smooth muscle cells. *Am. J. Physiol. Lung Cell Mol. Physiol.* **279**, L408–L412
46. Wood, J.G., Johnson, J.S., Mattioli, L.F. & Gonzalez, N.C. (1999) Systemic hypoxia promotes leukocyte-endothelial adherence via reactive oxidant generation. *J. Appl. Physiol.* **87**, 1734–1740
47. Fandrey, J., Frede, S. & Jelkmann, W. (1994) Role of hydrogen peroxide in hypoxia-induced erythropoietin production. *Biochem. J.* **303**, 507–510
48. Michelakis, E.D., Hampl, V., Nsair, A., Wu, X., Harry, G., Haromy, A., Gurtu, R. & Archer, S.L. (2002) Diversity in mitochondrial function explains differences in vascular oxygen sensing. *Circ. Res.* **90**, 1307–1315
49. Tarpey, M.M. & Fridovich, I. (2001) Methods of detection of vascular reactive species: nitric oxide, superoxide, hydrogen peroxide, and peroxynitrite. *Circ. Res.* **89**, 224–236
50. Pan, Y., Mansfield, K.D., Bertozzi, C.C., Rudenko, V., Chan, D.A., Giaccia, A.J. & Simon, M.C. (2007) Multiple factors affecting cellular redox status and energy metabolism modulate hypoxia-inducible factor prolyl hydroxylase activity *in vivo* and *in vitro*. *Mol. Cell. Biol.* **27**, 912–925
51. Gerald, D., Berra, E., Frapart, Y.M., Chan, D.A., Giaccia, A.J., Mansuy, D., Pouyssegur, J., Yaniv M. & Mechta-Grigoriou, F. (2004) JunD reduces tumor angiogenesis by protecting cells from oxidative stress. *Cell* **118**, 781–794
52. Turcotte, S., Desrosiers, R.R. & Beliveau, R (2003) HIF-1α mRNA and protein upregulation involves Rho GTPase expression during hypoxia in renal cell carcinoma. *J. Cell Sci.* **116,** 2247–2260
53. Mottet, D., Dumont, V., Deccache, Y., Demazy, C., Ninane, N., Raes, M. & Michiels, C. (2003) Regulation of hypoxia-inducible factor-1α protein level during hypoxic conditions by the phosphatidylinositol 3-kinase/Akt/glycogen synthase kinase 3β pathway in HepG2 cells. *J. Biol. Chem.* **278**, 31277–31285
54. Hirota, K. & Semenza, G.L. (2001) Rac1 activity is required for the activation of hypoxia-inducible factor 1. *J. Biol. Chem.* **276**, 21166–21172
55. Emerling, B.M., Platanias, L.C., Black, E., Nebreda, A.R., Davis, R.J. & Chandel N.S. (2005) Mitochondrial reactive oxygen species activation of p38 mitogen-activated protein kinase is required for hypoxia signaling. *Mol. Cell. Biol.* **25**, 4853–4862

3

Nitric oxide and hypoxia

Alexander Galkin, Annie Higgs
and Salvador Moncada[1]

The Wolfson Institute for Biomedical Research, The Cruciform Building, University College London, Gower Street, London WC1E 6BT, U.K.

Abstract

NO (nitric oxide) can affect mitochondrial function by interacting with the cytochrome *c* oxidase (complex IV) of the electron transport chain in a manner that is reversible and in competition with oxygen. Concentrations of NO too low to inhibit respiration can trigger cell defence response mechanisms involving reactive oxygen species and various signalling molecules such as nuclear factor κB and AMP kinase. Inhibition of mitochondrial respiration by NO at low oxygen concentrations can cause so-called metabolic hypoxia and divert oxygen towards other oxygen-dependent systems. Such a diversion reactivates prolyl hydroxylases and thus accounts for the prevention by NO of the stabilization of hypoxia-inducible transcription factor. In certain circumstances NO interacts with superoxide radical to form peroxynitrite, which can affect the action of key enzymes, such as mitochondrial complex I, by S-nitrosation. This chapter discusses the physiological and pathophysiological implications of the interactions of NO with the cytochrome *c* oxidase.

Introduction

Following the identification of EDRF (endothelium-derived relaxing factor) as NO (nitric oxide) and the subsequent demonstration of its biochemical synthesis from the semi-essential amino acid L-arginine, the term 'the L-arginine:NO pathway' was coined and its presence was confirmed in many different tissues

[1]*To whom correspondence should be addressed (email s.moncada@ucl.ac.uk).*

(see [1] for review). This was followed by the elucidation of a number of different biological roles for NO throughout the body. Many of these physiological actions of NO, such as its vasodilator, anti-aggregatory and neurotransmitter properties, result from activation of the soluble guanylate cyclase. In addition, NO can affect mitochondrial function by inhibiting the cytochrome *c* oxidase (complex IV) of the electron transport chain. The physiological and pathological consequences of such inhibition will be discussed in this chapter. We regret that, owing to editorial constraints, we have been unable to cite fully those who have contributed to this field of research.

Synthesis of NO in the cell

Three distinct isoforms of NOS (nitric oxide synthase), with different tissue localization, catalytic properties and regulatory mechanisms, have been identified. These are nNOS (neuronal NOS or NOS-I), eNOS (endothelial NOS or NOS-III) and iNOS (inducible NOS or NOS-II). The genes responsible for the expression of these proteins are known and the X-ray structure of the enzyme isoforms has been determined (see [2] for review). The enzyme catalyses the synthesis of NO from L-arginine and oxygen, as shown in reaction (1):

$$\text{L-Arginine}+2O_2+3/2\text{NADPH}+3/2H^+=\text{Citrulline}+2H_2O+3/2\text{NADP}+\text{NO} \quad (1)$$

In cells, all three isoforms occur as dimers, each monomer containing the following redox centres: FAD, two FMN molecules, iron protoporphyrin IX (haem) and tetrahydrobiopterin. Although the initial nomenclature was based on the tissue in which it was identified, it is now clear that each isomer has a widespread distribution (for review see [2]).

The existence of a mitochondrial isoform of the enzyme (termed mtNOS) has been proposed; however, to date this remains controversial [3]. It is now clear that there is no mtNOS localized to the mitochondrial matrix and that, most likely, a cytoplasmic isoform is associated with the mitochondrial membrane. Studies in endothelial cells have suggested that eNOS is localized at the cytoplasmic face of the outer mitochondrial membrane [4]. Another study has demonstrated a protein–protein interaction between mitochondrial nNOS and the cytochrome *c* oxidase in nervous tissue [5].

Generation of NO close to its site of action remains an intriguing possibility. In this context, the recent observation of NOS inhibitor-independent, nitrite-dependent, generation of NO by isolated yeast and rat liver mitochondria and by isolated cytochrome *c* oxidase at low oxygen concentrations [6] suggests an alternative mechanism whereby mitochondria themselves might generate NO from nitrite, initiating a hypoxia-signalling pathway. Furthermore, nitrite has been suggested to be an endocrine depot of NO, so that the nitrite reductase activity of deoxymyoglobin releases NO close to the mitochondria and thus regulates respiration via cytochrome *c* oxidase [7]. In addition, in the absence of oxygen, eNOS has been reported to catalyse the reduction of nitrite to NO [8]. Whether any of these alternative mechanisms involved in the generation of NO occur *in vivo* and/or play a role in the regulation of the cytochrome *c* oxidase or in cell signalling remains to be confirmed.

Interaction of NO with the mitochondrial respiratory chain

NO, a gas with a molecular mass and structure close to that of oxygen, has been known for more than 100 years to interact with transition metals and induce spectral changes in methaemoglobin by binding to the haem iron. It is not surprising, therefore, that the interaction of NO with cytochrome *c* oxidase, the terminal enzyme of the mitochondrial respiratory chain, was described long before the biological significance of NO was recognized in the 1980s. The inhibitory effect of NO on cellular cytochrome *c* oxidase was demonstrated in 1994 by several groups (for review see [9]). The inhibition is reversible and in competition with oxygen. The affinity of cytochrome *c* oxidase for NO is greater than that for oxygen, with a K_i of around 0.2 nM for competitive binding [10]. This indicates that interactions between oxygen and NO occur at physiological concentrations of both gases as has been demonstrated in a number of studies during the past few years. In vascular endothelial cells endogenous concentrations of NO modulate cell respiration in an oxygen-dependent manner [11] and in isolated canine cardiac muscle both exogenous and endogenous NO have been shown to reduce the consumption of oxygen [12]. The inhibition of cytochrome *c* oxidase by nanomolar concentrations of NO *in vitro* takes place via two different mechanisms so that NO can block the association of oxygen with the enzyme by binding to: (i) oxidized Cu_B^{2+}, reducing it to give a short lived Cu_B^{+}–NO^{+} intermediate with subsequent formation of nitrite (NO_2^{-}) ($k_{on}=10^5\ M^{-1}\cdot s^{-1}$); and (ii) the iron atom of the reduced cytochrome a_3 with formation of the a_3^{2+}–NO intermediate ($k_{on}=10^8\ M^{-1}\cdot s^{-1}$ and $k_{off}=0.004\ s^{-1}$ at 20°C). This is light-sensitive, strictly competitive with oxygen and mimicked by carbon monoxide. In both situations the net effect of NO interacting with cytochrome *c* oxidase is the reduction of the enzyme (for review see [10]).

We have demonstrated that inhibition of respiration by NO only occurs after the spare capacity of cytochrome *c* oxidase for oxygen is fully utilized by NO [13]. At lower concentrations NO reduces the electron transport chain without affecting its respiratory efficiency (see below). Since this latter effect of NO is the one associated with cell signalling [14], it is likely that the fine regulation of the behaviour of the enzyme by NO is indeed a physiological mechanism and occurs at concentrations of NO similar to those that activate the soluble guanylate cyclase enzyme, an interaction that is, incidentally, independent of oxygen concentration.

The interaction between NO and cytochrome *c* oxidase *in vivo* remains to be investigated fully; however, it has already been reported that administration of a NOS inhibitor or NO scavenger increases oxygen consumption in the heart, kidney, skeletal muscle and whole body during rest and exercise in conscious dogs [15], indicating the direct involvement of NO in regulation of mitochondrial respiration.

Free radical formation and NO

In 1971 the 'inhibitor-sensitive' generation of H_2O_2 (hydrogen peroxide) by metabolically efficient mitochondria was investigated and found to represent up to 2% of the total oxygen consumption in intact rat liver mitochondria [16]. Later it was shown that the initial ROS (reactive oxygen species) produced by the respiratory chain is superoxide radical ($O_2^{\bullet -}$) and that this is converted into H_2O_2, either non-enzymatically or by SOD (superoxide dismutase). H_2O_2 is a more stable, membrane-permeable molecule that can diffuse from the mitochondrial matrix to the cytoplasm and out of the cell and has been implicated in cell signalling (see below).

Respiratory complexes I and III are responsible for the one-electron reduction of molecular oxygen resulting in the production of $O_2^{\bullet -}$. At complex III the reaction occurs rapidly only if the physiological intramolecular electron transfer within the enzyme is disrupted by specific inhibitors (e.g. antimycin A). The unstable semiquinone located in the Q_o site is generally accepted to be the direct reductant of molecular oxygen (Figure 1). The production of $O_2^{\bullet -}$ by intact complex III is highly dependent on the mitochondrial membrane potential, as would be expected if it is produced at the Q_o site. Because of the absence of a clear mechanistic model of complex I there is less consensus on the site of generation of $O_2^{\bullet -}$ in this enzyme. However, recent data indicate that reduced flavin or its semiquinone anion is the most likely point of one-electron leak in complex I [17,18]. In contrast with complex III, none of the known redox centres of complex I are located in the

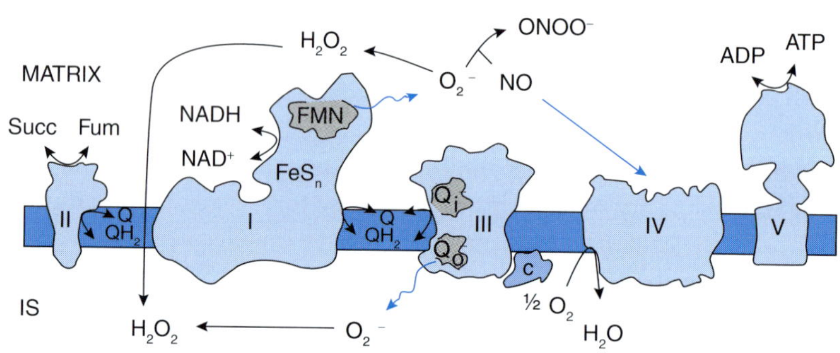

Figure 1. Scheme of the respiratory chain
Inhibition of cytochrome *c* oxidase (complex IV) by NO results in over-reduction of upstream redox centres and enhancement of superoxide radical ($O_2^{\bullet -}$) formation, most likely from complex I. The newly-formed $O_2^{\bullet -}$ is either efficiently dismutated by the mitochondrial Mn-SOD or, in certain circumstances, reacts with available NO to form $ONOO^-$. Succ, succinate; Fum, fumarate; IS, intracellular space.

membrane domain and even during uncoupled respiration all FeS clusters are in a reduced state. Therefore no significant enhancement of complex I-mediated $O_2^{\bullet -}$ production by changes in membrane potential would be expected, as has been shown experimentally [17,18]. The main feature of the generation of ROS by complex I is that the enzyme catalyses $O_2^{\bullet -}$ generation by both direct oxidation of NADH by ubiquinone and a reverse electron transfer reaction from ubiquinol to NAD^+ [18], releasing $O_2^{\bullet -}$ into the mitochondrial matrix. Therefore it is likely that *in situ*, in the absence of inhibitors, complex I rather than complex III is the major contributor to mitochondrial ROS generation during oxidation either of the substrates of NAD^+-dependent dehydrogenases or of succinate.

Reduction of the cytochrome *c* oxidase by NO at low oxygen concentrations has been shown in whole cells to be associated with the release of ROS and subsequent signalling for the activation of NF-κB (nuclear factor κB) [14] and AMP kinase [19]. It has also been demonstrated that increases in NO generation lead to reduction of cytochrome *c* oxidase at oxygen concentrations well above the hypoxic range, thus suggesting that this is not, strictly speaking, a hypoxic ROS-dependent signalling mechanism but that it depends on the relative ratio of NO to oxygen and is a stress response of wider biological significance [14]. Whether the release of ROS is due to the simple backlog of electrons that results from inhibition of the cytochrome *c* oxidase, or from the accumulation of reducing equivalents such as NADH and $FADH_2$ is not clear at present. The NO-induced enhanced reduction of the chain at higher oxygen concentrations leads to a situation that favours the interaction between electrons and oxygen; this is clearly different from the reduction of the chain at the very low oxygen concentrations of hypoxia [19].

At high concentrations exogenous NO has been shown to enhance the generation of H_2O_2 and $O_2^{\bullet -}$ in submitochondrial particles, rat heart mitochondria [20] and Langendorf preparations of isolated heart [21]. Whether this can be explained by the mechanisms above is not clear. In the presence of oxygen high NO concentrations may initiate a series of redox reactions with the consequent release of ROS or products of the reactions between NO and $O_2^{\bullet -}$. Full inhibition of respiration, with complete reduction of the cytochrome *c* oxidase and the entire respiratory chain, may not lead to the generation of ROS, as has indeed been shown using cyanide, which produces full inhibition of respiration without any significant increase in ROS production.

Interaction between NO and $O_2^{\bullet -}$ was identified early on during the quest for the identification of NO itself [1]. Later it was found that this interaction leads to the formation of $ONOO^-$ (peroxynitrite) and occurs at near-diffusion-limited rates ($6.7 \times 10^9\ M^{-1} \cdot s^{-1}$) that are even faster than the dismutation of $O_2^{\bullet -}$ by SOD. Direct measurements of $ONOO^-$ are not possible *in vivo*; however it has been measured indirectly by its ability to nitrosate proteins and these have been found in a number of pathologies in humans [22]. The source of $O_2^{\bullet -}$ for $ONOO^-$ generation has been widely speculated. Although the massive emission of highly reactive $O_2^{\bullet -}$ from mitochondria is improbable in the cell under physiological conditions, during pathological conditions the reaction between NO and $O_2^{\bullet -}$, coming from the mitochondria or from other sources (see below), is likely to occur. The formation in mitochondria of $ONOO^-$ from NO can take place only in the vicinity of the enzymes generating $O_2^{\bullet -}$ owing to the short half-life of $O_2^{\bullet -}$ and the high hydrophobicity of NO.

The precise conditions under which $ONOO^-$ is generated are unclear; however, long-term exposure of cells to NO leads to inhibition of cytochrome *c* oxidase that is followed by a gradual and persistent inhibition of complex I, concomitant with a reduction in the intracellular concentration of reduced glutathione [23]. We have suggested that this is likely to be the result of S-nitrosation of critical thiols in complex I since it can be reversed by exposing the cells to high-intensity light or by replenishing intracellular reduced glutathione. Thus, although NO regulates cell respiration physiologically by its action on cytochrome *c* oxidase, long-term exposure to high concentrations of NO leads to persistent inhibition of complex I and potentially to cell pathology. Whether the sequential inhibition of cytochrome *c* oxidase, then complex I, occurs in pathological conditions such as inflammation or degenerative disease is not known.

In spite of the above, there remain many uncertainties in relation to the extent to which mitochondria are a source of ROS under physiological and pathological conditions. Relatively little data have been published on mitochondrial production of H_2O_2 by metabolically competent mitochondria in the absence of inhibitors, and studies on the release of ROS from the mitochondrial respiratory chain in cells in real-time are still in their infancy [13,14]. Furthermore, although the respiratory chain is generally considered to be the main source of mitochondrial $O_2^{\bullet -}$ production, there are other enzymes in the matrix and outer mitochondrial membrane that are able to generate $O_2^{\bullet -}$ and/or H_2O_2. These include: DT-diaphorase, the hydrolipoamide dehydrogenase component of α-ketoglutarate dehydrogenase and pyruvate dehydrogenase complexes, monamine oxidase, dyhydroorotate dehydrogenase and α-glycerophosphate dehydrogenase (see [24] for review). To what extent any of these enzymes contributes to mitochondrial generation of ROS is unknown at present.

There is, in addition, little current data on the relative contribution of different cellular organelles and tissue-specific enzymes to the overall production of ROS in cells. Although the mitochondria and their respiratory chain enzymes, because of their strategic role in oxygen utilization, are likely to be the initiators of any generation of $O_2^{\bullet -}$ and subsequent signalling it is important to consider that even the earliest studies [25] found the mitochondria from rat liver to be responsible for only 15% of the total cellular H_2O_2 production, with the rest being generated by peroxisomes and microsomes. How this translates into the whole cell or the *in vivo* situation remains to be determined. Interestingly, mice lacking the gene encoding Mn-SOD, the enzyme located in the mitochondrial matrix, exhibit perinatal or early post-natal mortality which does appear to implicate the mitochondria as a major source and target of $O_2^{\bullet -}$ (see [26] for review). It is probable that other non-mitochondrial enzymes contribute to the total production of ROS in cells. So far, the activation of enzymes such as NADPH oxidases, xanthine oxidase and vascular cytochrome P450 has been implicated and substantial evidence now exists showing that the activity as well as the expression of these enzymes can be enhanced by pathological stimuli [27]. Furthermore, NOS has the capacity to generate $O_2^{\bullet -}$ under specific circumstances in which the enzyme becomes uncoupled [28]. The uncoupling of eNOS has been demonstrated in several pathological conditions such as diabetes, hyperc-holesterolaemia and hypertension. A more subtle mechanism for generation of $O_2^{\bullet -}$ by eNOS has been suggested to occur through the phosphorylation/dephosphorylation of a specific Thr^{495} which might act as a molecular switch between the generation of NO and that of $O_2^{\bullet -}$ [29]. Thus the whole topic of the release of ROS in cells is an evolving field with a great deal of research yet to be carried out. The fundamental questions that remain are to what extent the release of ROS is a physiological mechanism and how is this release modified during cell defence and pathophysiology?

S-nitrosation of proteins

S-nitrosation of thiol groups by the interaction of NO with cysteine has been suggested to be an important post-translational protein modification that plays a role in cell signalling [30]. Although S-nitrosation has been demonstrated in isolated enzymes *in vitro*, in cells and in organs, the mechanism by which it occurs remains unclear. Several possibilities that have been proposed include the reaction of NO itself with a pre-existing $RS^{\bullet}$ (thiyl) radical; with oxygen to form the electrophilic nitrosating agent N_2O_3, or with a transition metal to form the reactive electrophile NO^+. Peroxynitrite has also been suggested to oxidize cysteine residues to form S-nitrosoproteins. Since S-nitrosation appears to be targeted to specific sites on particular proteins some authors have suggested the existence of certain cysteine-containing amino acid consensus motifs in proteins promoting the nitrosating reaction. The pK_A of the cysteine group (from 8 to 10) may contribute to its reactivity, as may the proximity of the target protein to the NO generator (for review see [31]).

Since gaseous NO is a very poor nitrosating agent and has a relatively short half-life in aqueous medium, it is likely that transnitrosation from one thiol group to another could account for some nitrosation reactions. Glutathione, the most abundant thiol-containing peptide in cells, has been suggested to be the potential main depot of the nitrosothiol available for transnitrosation. Some enzymes (thioredoxin reductase, glutathione-dependent formaldehyde dehydrogenase, SOD, glutathione peroxidase) have been shown to catalyse the reductive cleavage of S-nitrosoglutathione, as does protein disulfide isomerase. Transnitrosation between S-nitrosated proteins and S-nitrosothiols may also facilitate the transfer of NO from extracellular to intracellular thiol pools (for reviews see [30,31]).

During the last decade many proteins have been reported to undergo regulatory modification by S-nitrosation: these include serum albumin, glyceraldehyde-3-phosphate dehydrogenase, creatine kinase, caspase, actin, haemoglobin, ryanodine-sensitive calcium release channels, methionine adenosyl transferase and NMDA receptors (for review see [30]). The list of NO-susceptible proteins, including receptors, enzymes and transcription factors, is extending continually as new examples of regulatory S-nitrosation are added. Since S-nitrosation of a protein is likely to alter its activity, such a modification may have biological relevance. It has been shown, for example, that the NO-dependent inhibition of NOS previously demonstrated [32] may be due to S-nitrosation (see [31]). Also, the anti-apoptotic action of NO has been suggested to be due to S-nitrosation of caspase-3 at Cys^{163}, resulting in enzyme inactivation. This S-nitrosation of the caspase has been shown to occur via transnitrosation by thioredoxin-Cys^{73}-SNO in a reaction that occurs 100 times faster than that of S-nitrosoglutathione and is specific for Cys^{163} of caspase [33].

As discussed above, we found that persistent inhibition of complex I following long-term inhibition of cytochrome *c* oxidase by NO was most probably due to S-nitrosation of this enzyme [23]. This has been confirmed by other research groups [34,35]. Our work [36] and that of others [37] has also indicated that mitochondria are crucial for the formation of S-nitrosoproteins, possibly because of their capacity to generate $O_2^{\bullet-}$ and, as a result, $ONOO^-$. More recently it has been shown that, in addition, S-nitrosation of complex I increases the production of H_2O_2 by mitochondria [38].

It is still not clear whether these processes are linked with physiological signalling responses of the cell or are the beginning of an early response to oxidative stress and pathology. It has been claimed that S-nitrosation of complex I is protective in rat hearts subjected to ischaemia reperfusion injury since the S-nitrosated enzyme can degrade and release NO [34]. Furthermore, altered blood levels of S-nitrosated proteins are associated with impaired clinical outcome in patients with cardiovascular disease [39]. Conversely, inhibition of complex I by S-nitrosation has been implicated in a number of pathologies, including Huntington's, Alzheimer's and Parkinson's diseases [34]. Furthermore, decreased respiratory chain complex I activity has been reported in skeletal muscle biopsies taken from critically ill patients in septic shock [40].

Hypoxia, diversion of oxygen and NO

The intracellular concentration and the dynamics of oxygen redistribution between mitochondria and cytosol are likely to be tissue-specific. Hypoxia, a state of oxygen deficiency which is sufficient to cause an impairment of function, initiates a sequence of cellular responses. A major signalling pathway is via the HIFs (hypoxia-inducible factors). HIF is a heterodimer consisting of two subunits, HIF-1α and HIF-1β. HIF-1β is constitutively expressed and its concentration is not affected by changes in cellular PO_2 (partial pressure of oxygen). HIF-1α, on the other hand, is subjected to degradation under normal conditions. However, in response to hypoxia, its degradation is inhibited and it dimerizes with HIF-1β, resulting in active HIF-1 (see [41] for review). This leads to binding of HIF-1 to DNA and to up-regulation of the expression of specific genes, including those involved in glucose metabolism, angiogenesis and haematopoiesis.

The HIF-1α subunit is continuously degraded by the ubiquitin–proteasome system after hydroxylation of the amino acids Pro^{564} and/or Pro^{402} (see other chapters in this volume) and subsequent binding of the pVHL (von Hippel-Lindau protein). HIF prolyl hydroxylation is carried out by a family of PHDs (prolyl hydroxylases) whose action is impaired in anoxic or low oxygen conditions, leading to decreased pVHL-ubiquitination and increased HIF-1α stability. Interestingly, a defect in the VHL gene is associated with a variety of cancers, particularly renal cancer, since over-accumulation of HIF results in overproduction of proteins important in this disease such as vascular endothelial growth factor [42].

Following the demonstration that high concentrations of H_2O_2 induce stabilization of HIF-1α, Schumacker and co-workers [43] proposed that HIF-1α is stabilized by ROS generated by mitochondria. Evidence for this comes from several observations, including the fact that loss of cytochrome *c* oxidase in cells [44], or the use of a mitochondrial-targeted antioxidant [45], prevent HIF-1α stabilization at low oxygen concentrations. This remains controversial, however, since a great deal of evidence suggests that during hypoxia there is actually a drop in ROS production [26] and that this initiates the signalling pathway leading to stabilization of HIF-1α. Observations that enzymatically generated $O_2^{\bullet-}$ reduces, and SOD increases, HIF-1α levels also suggest an inhibitory role for ROS [46].

The involvement of a free radical mechanism in the stabilization of HIF-1α is supported by some of our experiments in which it is stabilized by NO at ambient oxygen concentrations in a mitochondrial-independent manner [47]. This occurs using high concentrations of NO, which probably stabilize HIF-1α as a result of $ONOO^-$ generation following the interaction between NO and $O_2^{\bullet-}$. Furthermore in some cancer cells we have demonstrated stabilization of HIF-1α by a free radical mechanism involving NO [48]. These experiments are in agreement with data in the literature showing that S-nitrosation of either the PHDs or the HIF-1α molecule can affect its degradation [49]. Moreover, an interaction of free radicals with iron in the PHDs has been suggested as a possible mechanism for their inactivation [50]. Whether, therefore, at some point in the cellular response to decreasing oxygen concentrations, a mitochondrial-dependent formation of ROS synergizes with a progressive inactivation of the PHDs to stabilize HIF-1α remains an interesting possibility. Indeed, there is evidence for a strong synergism between ROS- and hypoxia-induced stabilization of HIF-1α [47,48].

The term 'metabolic hypoxia' was introduced by Moncada and Erusalimsky [9] to describe the situation in which oxygen is available for mitochondria but the organelle is unable to use it due to inhibition of the respiratory chain by NO. Metabolic hypoxia is likely to have two phases, the first due to the reversible inhibition of cytochrome *c* oxidase and the second to the more persistent inhibition of complex I and other key enzymes [23,51]. Using a system in which a human cell line is transfected so that it generates finely-controlled amounts of NO we have shown that low concentrations cause a rapid decrease in HIF-1α stabilized by exposure of the cells to 3% oxygen [47]. This effect of NO is dependent on the inhibition of mitochondrial respiration since it is mimicked by other respiratory chain inhibitors (rotenone, myxothiazol and cyanide). In addition, the destabilizing effect on HIF-1α of NO and other inhibitors of mitochondrial respiration in those experiments was independent of ROS since it was neither prevented nor enhanced by antioxidants [52]. The effect of inhibition of respiration on HIF-1α stability is dependent on the reactivation of PHDs since we showed that, upon inhibition of mitochondrial respiration in hypoxia, oxygen is redistributed towards non-respiratory oxygen-dependent targets such as PHD so that they do not register hypoxia [52]. These results have been confirmed in other systems [53]. A clear demonstration of oxygen redistribution is the fact that cells lacking mitochondrial DNA, and therefore unable to utilize oxygen, stabilize HIF-1α at lower concentrations than those required by the parent cell [47]. Thus it is possible that NO-dependent diversion of oxygen may be a mechanism whereby enzymes whose activities are reduced in hypoxia may be reactivated, and during pathology it may be the origin of metabolic changes that occur in inflammation and degenerative diseases.

NO and mitochondrial biogenesis

We have demonstrated that NO activates mitochondrial biogenesis, i.e. mitochondrial proliferation and activation, in various types of cells including brown adipose tissues and HeLa cells [54]. This effect of NO, which has been confirmed by others in rat skeletal muscle [55], is dependent on activation of the guanylate cyclase and is mediated by the induction of PGC-1α (peroxisome proliferator-activated receptor γ coactivator 1α), the main regulator of mitochondrial biogenesis and oxidative phosphorylation. We have also shown that mitochondrial biogenesis is reduced in $eNOS^{-/-}$ mice following either exposure to cold or calorie restriction, thus providing further evidence that a NO/cGMP-dependent pathway plays a significant role in the control of mitochondrial biogenesis and the body energy balance [56].

We have shown more recently that in several animal models of obesity eNOS expression and mitochondrial biogenesis are reduced in a manner that

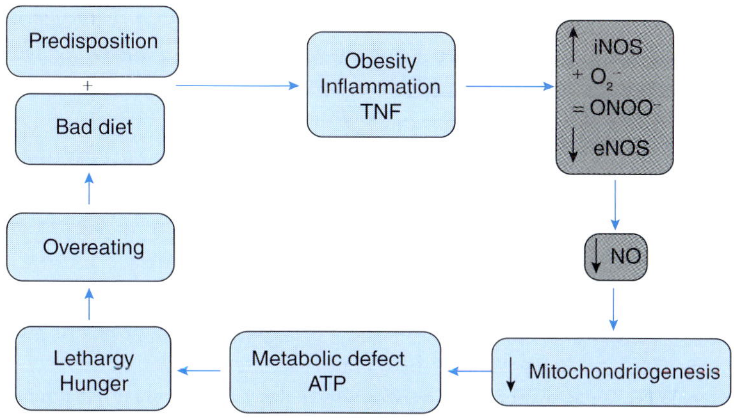

Figure 2. Schematic diagram of the vicious cycle of obesity
It is likely that the steps shown in the grey boxes play a significant role in the pathophysiology of this condition.

can be reversed by preventing the signalling of TNFα (tumour necrosis factor α) [57]. This led us to suggest that reduced mitochondriogenesis in obesity might be the result of the inflammatory reaction, now known to co-exist with adiposity, leading to a switch from eNOS to iNOS activity. We have also suggested that this change may lead to a decrease in NO in favour of NO adducts which will promote further inflammation. Thus it appears as if there is a vicious cycle in obesity in which a genetic predisposition combines with an unhealthy diet to promote obesity and inflammation. Cytokines such as TNFα reduce the availability of NO generated from eNOS, while promoting the generation of iNOS-derived $ONOO^-$, thus reducing mitochondriogenesis. This leads to a lack of ATP production and the resulting hunger and lethargy encourages overeating and greater obesity (Figure 2).

Conclusions

NO plays different roles in multiple biological systems, suggesting a significant regulatory function in cell physiology and also in cell adaptation to stress leading to pathophysiology. These roles are mediated by the well-known activation of the soluble guanylate cyclase, as well as by competing with oxygen at the cytochrome *c* oxidase and most probably via a wide variety of other redox interactions which are at present being actively investigated.

Summary

- *NO interacts with cytochrome c oxidase in a manner that is competitive with oxygen.*
- *This interaction can trigger cell defence response mechanisms involving ROS and various signalling molecules.*
- *In certain circumstances NO interacts with $O_2^{\bullet-}$ to form $ONOO^-$, which can affect the action of key enzymes such as complex I by S-nitrosation.*

- *Inhibition of mitochondrial respiration by NO at low oxygen concentrations can divert oxygen towards other oxygen-dependent systems. Such a diversion reactivates PHDs and thus accounts for the prevention by NO of the stabilization of HIF.*
- *NO promotes mitochondriogenesis by a mechanism involving soluble guanylate cyclase rather than cytochrome c oxidase*

References

1. Moncada, S. (2006) Adventures in vascular biology: a tale of two mediators. *Phil. Trans. Roy. Soc. B.* **361**, 735–759
2. Alderton, W.K., Cooper, C.E. & Knowles, R.G. (2001) Nitric oxide synthases: structure, function and inhibition. *Biochem. J.* **357**, 593–615
3. Brookes, P.S. (2004) Mitochondrial nitric oxide synthase. *Mitochondrion* **3**, 187–204
4. Gao, S., Chen, J., Brodsky, S.V., Huang, H., Adler, S., Lee, J.H., Dhadwal, N., Cohen-Gould, L., Gross, S.S. & Goligorsky, M.S. (2004) Docking of endothelial nitric oxide synthase (eNOS) to the mitochondrial outer membrane: a pentabasic amino acid sequence in the autoinhibitory domain of eNOS targets a proteinase K-cleavable peptide on the cytoplasmic face of mitochondria. *J. Biol. Chem.* **279**, 15968–15974
5. Persichini, T., Mazzone, V., Polticelli, F., Moreno, S., Venturini, G., Clementi, E. & Colasanti, M. (2005) Mitochondrial type I nitric oxide synthase physically interacts with cytochrome *c* oxidase. *Neurosci. Lett.* **384**, 254–259
6. Castello, P.R., David, P.S., McClure, T., Crook, Z. & Poyton, R.O. (2006) Mitochondrial cytochrome oxidase produces nitric oxide under hypoxic conditions: implications for oxygen sensing and hypoxic signaling in eukaryotes. *Cell Metab.* **3**, 277–287
7. Shiva, S., Huang, Z., Grubina, R., Sun, J., Ringwood, L.A., Macarthur, P.H., Xu, X., Murphy, E., Darley-Usmar, V.M. & Gladwin, M.T. (2007) Deoxymyoglobin is a nitrite reductase that generates nitric oxide and regulates mitochondrial respiration. *Circ. Res.* **100**, 654–661
8. Vanin, A.F., Bevers, L.M., Slama-Schwok, A. & van Faassen, E.E. (2007) Nitric oxide synthase reduces nitrite to NO under anoxia. *Cell Mol. Life Sci.* **64**, 96–103
9. Moncada, S. & Erusalimsky, J.D. (2002) Does nitric oxide modulate mitochondrial energy generation and apoptosis? *Nat. Rev. Mol. Cell Biol.* **3**, 214–220
10. Cooper, C.E. & Giulivi, C. (2007) Nitric oxide regulation of mitochondrial oxygen consumption II: molecular mechanism and tissue physiology. *Am. J. Physiol. Cell Physiol.* **292**, C1993–C2003
11. Clementi, E., Brown, G.C., Foxwell, N. & Moncada, S. (1999) On the mechanism by which vascular endothelial cells regulate their oxygen consumption. Proc. Natl. Acad. Sci. U.S.A. **96**, 1559–1562
12. Xie, Y.W., Shen, W., Zhao, G., Xu, X., Wolin, M.S. & Hintze, T.H. (1996) Role of endothelium-derived nitric oxide in the modulation of canine myocardial mitochondrial respiration in vitro. Implications for the development of heart failure. *Circ. Res.* **79**, 381–387
13. Palacios-Callender, M., Hollis, V., Frakich, N., Mateo, J. & Moncada, S. (2007) Cytochrome *c* oxidase maintains mitochondrial respiration during partial inhibition by nitric oxide. *J. Cell Sci.* **120**, 160–165
14. Palacios-Callender, M., Quintero, M., Hollis, V.S., Springett, R.J. & Moncada, S. (2004) Endogenous NO regulates superoxide production at low oxygen concentrations by modifying the redox state of cytochrome *c* oxidase. *Proc. Natl. Acad. Sci. U.S.A.* **101**, 7630–7635
15. Shen, W., Xu, X., Ochoa, M., Zhao, G., Bernstein, R.D., Forfia, P. & Hintze, T.H. (2000) Endogenous nitric oxide in the control of skeletal muscle oxygen extraction during exercise. *Acta Physiol. Scand.* **168**, 675–686
16. Boveris, A. & Chance, B. (1973) The mitochondrial generation of hydrogen peroxide. General properties and effect of hyperbaric oxygen. *Biochem. J.* **134**, 707–716.

17. Galkin, A. & Brandt, U. (2005) Superoxide radical formation by pure complex I (NADH:ubiquinone oxidoreductase) from *Yarrowia lipolytica*. *J. Biol. Chem.* **280**, 30129–30135
18. Grivennikova, V.G. & Vinogradov, A.D. (2006) Generation of superoxide by the mitochondrial complex I. *Biochim. Biophys. Acta* **1757**, 553–561
19. Quintero, M., Colombo, S.L., Godfrey, A. & Moncada, S. (2006) Mitochondria as signaling organelles in the vascular endothelium. *Proc. Natl. Acad. Sci. U.S.A.* **103**, 5379–5384
20. Poderoso, J.J., Carreras, M.C., Lisdero, C., Riobo, N., Schopfer, F. & Boveris, A. (1996) Nitric oxide inhibits electron transfer and increases superoxide radical production in rat heart mitochondria and submitochondrial particles. *Arch. Biochem. Biophys.* **328**, 85–92
21. Poderoso, J.J., Peralta, J.G., Lisdero, C.L., Carreras, M.C., Radisic, M., Schopfer, F., Cadenas, E. & Boveris, A. (1998) Nitric oxide regulates oxygen uptake and hydrogen peroxide release by the isolated beating rat heart. *Am. J. Physiol.* **274**, C112–C119
22. Greenacre, S.A. & Ischiropoulos, H. (2001) Tyrosine nitration: localization, quantification, consequences for protein function and signal transduction. *Free Radical Res.* **34**, 541–581
23. Clementi, E., Brown, G.C., Feelisch, M. & Moncada, S. (1998) Persistent inhibition of cell respiration by nitric oxide: crucial role of S-nitrosylation of mitochondrial complex I and protective action of glutathione. *Proc. Natl. Acad. Sci. U.S.A.* **95**, 7631–7636
24. Andreyev, A.Y., Kushnareva, Y.E. & Starkov, A.A. (2005) Mitochondrial metabolism of reactive oxygen species. *Biochemistry* **70**, 200–214
25. Boveris, A., Oshino, N. & Chance, B. (1972) The cellular production of hydrogen peroxide. *Biochem. J.* **128**, 617–630
26. Chandel, N.S. & Budinger, G.R. (2007) The cellular basis for diverse responses to oxygen. *Free Radical Biol. Med.* **42**, 165–174
27. Mueller, C.F., Laude, K., McNally, J.S. & Harrison, D.G. (2005) Redox mechanisms in blood vessels. *Arterioscler. Thromb. Vasc. Biol.* **25**, 274–278
28. Stuehr, D., Pou, S. & Rosen, G.M. (2001) Oxygen reduction by nitric oxide synthases. *J. Biol. Chem.* **276**, 14533–14536
29. Lin, M.I., Fulton, D., Babbitt, R., Fleming, I., Busse, R., Pritchard, Jr, K.A. & Sessa, W.C. (2003) Phosphorylation of threonine 497 in endothelial nitric oxide synthase coordinates the coupling of L-arginine metabolism to efficient nitric oxide production. *J. Biol. Chem.* **278**, 44719–44726
30. Hess, D.T., Matsumoto, A., Kim, S.O., Marshall, H.E. & Stamler, J.S. (2005) Protein S-nitrosylation: purview and parameters. *Nat. Rev. Mol. Cell Biol.* **6**, 150–166
31. Handy, D.E. & Loscalzo, J. (2006) Nitric oxide and posttranslational modification of the vascular proteome: S-nitrosation of reactive thiols. *Arterioscler. Thromb. Vasc. Biol.* **26**, 1207–1214
32. Assreuy, J., Cunha, F.Q., Liew, F.Y. & Moncada, S. (1993) Feedback inhibition of nitric oxide synthase activity by nitric oxide. *Br. J. Pharmacol.* **108**, 833–837
33. Mitchell, D.A. & Marletta, M.A. (2005) Thioredoxin catalyzes the S-nitrosation of the caspase-3 active site cysteine. *Nat. Chem. Biol.* **1**, 154–158
34. Burwell, L.S., Nadtochiy, S.M., Tompkins, A.J., Young, S. & Brookes, P.S. (2006) Direct evidence for S-nitrosation of mitochondrial complex I. *Biochem. J.* **394**, 627–634
35. Dahm, C.C., Moore, K. & Murphy, M.P. (2006) Persistent S-nitrosation of complex I and other mitochondrial membrane proteins by S-nitrosothiols but not nitric oxide or peroxynitrite: implications for the interaction of nitric oxide with mitochondria. *J. Biol. Chem.* **281**, 10056–10065
36. Frost, M.T., Wang, Q., Moncada, S. & Singer, M. (2005) Hypoxia accelerates nitric oxide-dependent inhibition of mitochondrial complex I in activated macrophages. *Am. J. Physiol. Regul. Integr. Comp. Physiol.* **288**, R394–R400
37. Yang, Y. & Loscalzo, J. (2005) S-nitrosoprotein formation and localization in endothelial cells. *Proc. Natl. Acad. Sci. U.S.A.* **102**, 117–122
38. Borutaite, V. & Brown, G.C. (2006) S-nitrosothiol inhibition of mitochondrial complex I causes a reversible increase in mitochondrial hydrogen peroxide production. *Biochim. Biophys. Acta* **1757**, 562–566
39. Massy, Z.A., Fumeron, C., Borderie, D., Tuppin, P., Nguyen-Khoa, T., Benoit, M.O., Jacquot, C., Buisson, C., Drueke, T.B., Ekindjian, O.G. et al. (2004) Increased plasma S-nitrosothiol concen-

trations predict cardiovascular outcomes among patients with end-stage renal disease: a prospective study. *J. Am. Soc. Nephrol.* **5**, 470–476

40. Brealey, D., Brand, M., Hargreaves, I., Heales, S., Land, J., Smolenski, R., Davies, N.A,, Cooper, C.E. & Singer, M. (2002) Association between mitochondrial dysfunction and severity and outcome of septic shock. **360**, 219–223
41. Semenza, G.L. (2006) Regulation of physiological responses to continuous and intermittent hypoxia by hypoxia-inducible factor 1. *Exp. Physiol.* **91**, 803–806
42. Gnarra, J.R., Tory, K., Weng, Y., Schmidt, L., Wei, M.H., Li, H., Latif, F., Liu, S., Chen, F., Duh, F.M. et al. (1994) Mutations of the VHL tumour suppressor gene in renal carcinoma. *Nat. Genet.* **7**, 85–90
43. Chandel, N.S., McClintock, D.S., Feliciano, C.E., Wood, T.M., Melendez, J.A., Rodriguez, A.M. & Schumacker, P.T. (2000) Reactive oxygen species generated at mitochondrial complex III stabilize hypoxia-inducible factor-1α during hypoxia. A mechanism of O_2 sensing. *J. Biol. Chem.* **27**, 25130–25138
44. Mansfield, K.D., Guzy, R.D., Pan, Y., Young, R.M., Cash, T.P., Schumacker, P.T. & Simon, M.C. (2005) Mitochondrial dysfunction resulting from loss of cytochrome *c* impairs cellular oxygen sensing and hypoxic HIF-α activation. Cell Metab. **1**, 393–399
45. Sanjuan-Pla, A., Cervera, A.M., Apostolova, N., Garcia-Bou, R., Victor, V.M., Murphy, M.P. & McCreath, K.J. (2005) A targeted antioxidant reveals the importance of mitochondrial reactive oxygen species in the hypoxic signaling of HIF-α. *FEBS Lett.* **579**, 2669–2674
46. Wellman, T.L., Jenkins, J., Penar, P.L., Tranmer, B., Zahr, R. & Lounsbury, K.M. (2004) Nitric oxide and reactive oxygen species exert opposing effects on the stability of hypoxia-inducible factor-1α (HIF-1α) in explants of human pial arteries. *FASEB J.* **18**, 379–381
47. Mateo, J., Garcia-Lecea, M., Cadenas, S., Hernandez, C. & Moncada, S. (2003) Regulation of hypoxia-inducible factor-1α by nitric oxide through mitochondria-dependent and -independent pathways. *Biochem. J.* **376**, 537–544
48. Quintero, M., Brennan, P.A., Thomas, G.J. & Moncada, S. (2006) Nitric oxide is a factor in the stabilization of hypoxia-inducible factor-1α in cancer: role of free radical formation. *Cancer Res.* **66**, 770–774
49. Metzen, E., Zhou, J., Jelkmann, W., Fandrey, J. & Brune, B. (2003) Nitric oxide impairs normoxic degradation of HIF-1alpha by inhibition of prolyl hydroxylases. *Mol. Biol. Cell* **14**, 3470–3481
50. Gerald, D., Berra, E., Frapart, Y.M., Chan, D.A., Giaccia, A.J., Mansuy, D., Pouyssegur, J., Yaniv, M. & Mechta-Grigoriou, F. (2004) JunD reduces tumor angiogenesis by protecting cells from oxidative stress. *Cell* **118**, 781–794
51. Beltran, B., Mathur, A., Duchen, M.R., Erusalimsky, J.D. & Moncada, S. (2000) The effect of nitric oxide on cell respiration: a key to understanding its role in cell survival or death. *Proc. Natl. Acad. Sci. U.S.A.* **97**, 14602–14607
52. Hagen, T., Taylor, C.T., Lam, F. & Moncada, S. (2003) Redistribution of intracellular oxygen in hypoxia by nitric oxide: effect on HIF1α. *Science* **302**, 1975–1978
53. Doege, K., Heine, S., Jensen, I., Jelkmann, W. & Metzen, E. (2005) Inhibition of mitochondrial respiration elevates oxygen concentration but leaves regulation of hypoxia-inducible factor (HIF) intact. *Blood* **106**, 2311–2317
54. Nisoli, E., Clementi, E., Paolucci, C., Cozzi, V., Tonello, C., Sciorati, C., Bracale, R., Valerio, A., Francolini, M., Moncada, S. & Carruba, M.O. (2003) Mitochondrial biogenesis in mammals: the role of endogenous nitric oxide. *Science* **299**, 896–899
55. Wadley, G.D. & McConell, G.K. (2007) Effect of nitric oxide synthase inhibition on mitochondrial biogenesis in rat skeletal muscle. *J. Appl. Physiol.* 102, 314–320
56. Nisoli, E., Tonello, C., Cardile, A., Cozzi, V., Bracale, R., Tedesco, L., Falcone, S., Valerio, A., Cantoni, O., Clementi, E., Moncada, S. & Carruba, M.O. (2005) Calorie restriction promotes mitochondrial biogenesis by inducing the expression of eNOS. *Science* **310**, 314–317
57. Valerio, A., Cardile, A., Cozzi, V., Bracale, R., Tedesco, L., Pisconti, A., Palomba, L., Cantoni, O., Clementi, E., Moncada, S., Carruba, M.O. & Nisoli, E. (2006) TNFα downregulates eNOS expression and mitochondrial biogenesis in fat and muscle of obese rodents. *J. Clin. Invest.* **116**, 2791–2798

4

Sensing hypoxia in the carotid body: from stimulus to response

Prem Kumar[1]

Department of Physiology, The Medical School, University of Birmingham, Birmingham B15 2TT, U.K.

Abstract

The carotid body is a peripheral sensory organ that can transduce modest falls in the arterial PO_2 (partial pressure of oxygen) into a neural signal that provides the afferent limb of a set of stereotypic cardiorespiratory reflexes that are graded according to the intensity of the stimulus. The stimulus sensed is tissue PO_2 and this can be estimated to be around 50 mmHg during arterial normoxia, falling to between 10–40 mmHg during hypoxia. The chemoafferent hypoxia stimulus-response curve is exponential, rising in discharge frequency with falling PO_2, and with no absolute threshold apparent in hyperoxia. Although the oxygen sensor has not been definitely identified, it is believed to reside within type I cells of the carotid body, and presently two major hypotheses have been put forward to account for the sensing mechanism. The first relies upon alterations in the cell energy status that is sensed by the cytosolic enzyme AMPK (AMP-activated protein kinase) subsequent to hypoxia-induced increases in the cellular AMP/ATP ratio during hypoxia. AMPK is localized close to the plasma membrane and its activation can inhibit both large conductance, calcium-activated potassium (BK) and background, TASK-like potassium channels, inducing membrane depolarization, voltage-gated calcium entry and neurosecretion of a range of transmitter and modulator substances,

[1]*To whom correspondence should be addressed (email p.kumar@bham.ac.uk).*

including catecholamines, ATP and acetylcholine. The alternative hypothesis considers a role for haemoxygenase-2, which uses oxygen as a substrate and may act to gate an associated BK channel through the action of its products, carbon monoxide and possibly haem. It is likely however, that these and other hypotheses of oxygen transduction are not mutually exclusive and that each plays a role, via its own particular sensitivity, in shaping the full response of this organ between hyperoxia and anoxia.

Introduction

Type I cells of the mammalian carotid body sense hypoxia; that is, they have the ability to detect and convert, within seconds, physiological decreases in the PO_2 (partial pressure of oxygen) of arterial blood into calcium-dependent neurosecretion that results in coded neural discharge in the post-synaptically located carotid sinus nerve afferents. This discharge is appropriately graded for the initiation of remedial cardiorespiratory reflexes that act to reduce the impact of systemic hypoxia. Carotid body hypoxia chemotransduction thus differs from oxygen sensing in many other tissues and cells of the body as it demonstrates a rapidly responding, low affinity sensing system, i.e. acting without alteration in protein expression levels and at degrees of PO_2 that should be without compromise to cellular metabolism, for the initiation of reflex responses to hypoxaemia. Although hypoxia is believed to be the key excitatory stimulus of the carotid body, it is important to remember that what is being sensed is a lack of oxygen. Put another way, hypoxia must remove or modify some component of cellular physiology that normally occurs in the presence of adequate levels of molecular oxygen. Although a consensus is emerging that places the inhibition of specific potassium channels located within the plasmalemmal membrane of type I cells of the carotid body as pivotal to the sensing mechanism, no single sensor for a lack of oxygen has yet been convincingly defined and it may be that a number of sensors, each with different oxygen affinities and downstream targets, co-exist to provide the graded response that is known to occur throughout the wide range of PO_2 between hyperoxia and anoxia. This chapter aims to provide an introduction and general overview of the main features of carotid body chemoreception, from the nature of the stimulus, to the generation of action potentials in response to acute hypoxia and through to the molecular mechanism involved in stimulus transduction while attempting to highlight where consensus has yet to be established and possible reasons for this.

Tissue PO_2 is the adequate stimulus of the carotid body

The carotid bodies are small (<100 μg wet weight in the rat) sensory organs located bilaterally in the vicinity of the bifurcation of the common carotid arteries where they branch into the internal and external carotid arteries in the neck of mammals (Figure 1). They receive their sensory, afferent innervation from the carotid sinus nerve, a branch of the glossopharyngeal (IXth cranial)

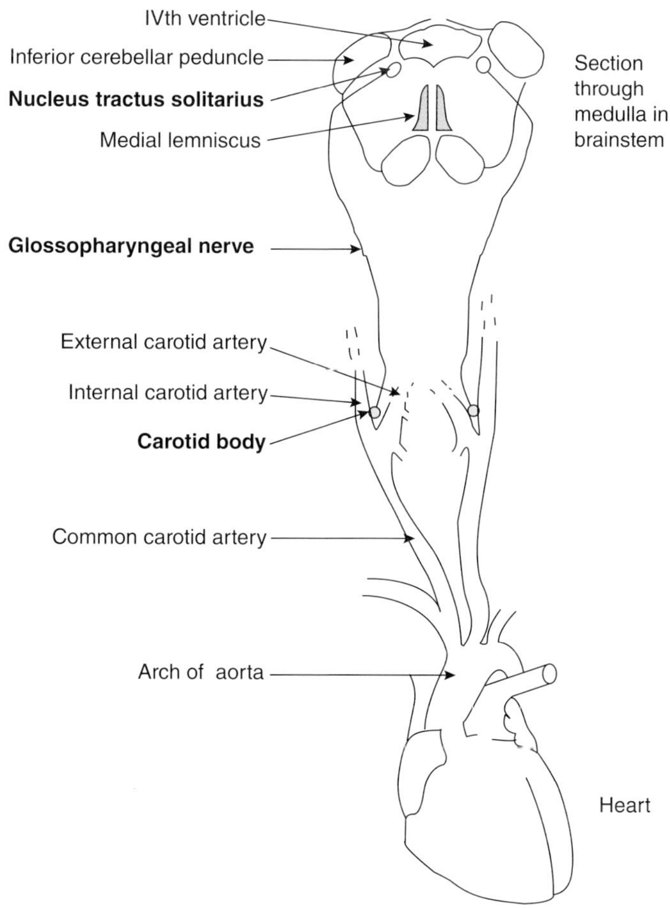

Figure 1. Anatomical location of the carotid bodies
The carotid bodies are located, bilaterally, at the bifurcation of the common carotid arteries into the internal and externals carotid arteries. Although their precise location in the bifurcation can vary slightly between species, in humans they are more often associated with the external rather than the internal carotid artery. The afferent supply of the carotid body is via the carotid sinus branch (not shown due to scale) of the glossopharyngeal (IXth cranial) nerve which enables chemoreceptor sensory information to be transmitted to the NTS in the medulla of the brainstem, from where the information can be integrated into appropriate reflex responses mediated through respiratory motorneurones and the autonomic nervous system.

nerve that has its cell bodies located in the petrosal ganglion. These afferent fibres project from the NTS (nucleus tractus solitarius) of the medulla where they mediate cardiorespiratory and autonomic reflex responses. The carotid body also receives a sympathetic, efferent innervation from the closely located superior cervical ganglion and an additional efferent innervation from parasympathetic fibres originating from the vagus (Xth cranial) nerve. The organ is protected by a connective tissue capsule and is highly vascular, receiving a blood flow, relative to its size, of 10–15 times that received by the brain. In addition to a capillary supply consisting of both typical (approx. 7–8 μm) diameter vessels and wider (maximum 18 μm) diameter penetrating sinusoid capillaries that make more intimate contact with the type I cells, there is also anatomical evidence that the carotid body possesses arterio-venous shunts, thus providing various routes for blood to flow through the organ. Its efferent innervation may act principally to shape the afferent response of the organ by modulation of this blood flow and hence the stimulus intensity in the vicinity of the sensory cells. An extensive anatomical description has been described elsewhere [1]. The high blood flow (approx. 2000 ml per 100 $g^{-1}\cdot min^{-1}$, at normal blood gases and blood pressure) ensures a very low arterio-venous oxygen difference in normoxia and provides sufficient oxygen for metabolic demands during periods of increased activity. Technical difficulties confound accurate measurements of oxygen consumption, but a value of between 1–2 ml per 100 $g^{-1}\cdot min^{-1}$ during normoxia is a best estimate. In accordance with its sensory function, on lowering PO_2, the carotid body markedly increases its glucose consumption and oxidation and an increase in the rate of oxygen disappearance has been positively correlated to sinus nerve afferent discharge [2]. Thus oxygen is needed to provide the energy necessary to sustain the chemosensory responses during hypoxia but, at some point, with increasing hypoxia, failure to sustain aerobic metabolism must occur.

Each carotid body consists of thousands of 8–12 μm diameter, spherical type I cells arranged in clustered islets of 3–5 cells (glomeruli) within a lobular structure. Each cluster is closely associated with a single, elongated type II cell, blood vessels and an afferent nerve supply. Numerous studies have identified the type I cell as the oxygen-sensing element with the type II cell allocated a supportive, glial-like role. Type II cells do not appear to make synaptic contact with afferent nerve fibres but type I cells do. In addition, type I cells are connected to each other via chemical and electrical gap junctions and the consequences of losing these communication pathways during cell isolation procedures may account for some of the differences noted between different preparations. The question of a heterogeneous population of type I cells has been raised, but this is not really considered in the present hypotheses of chemotransduction, and there is no evidence to suggest otherwise than all afferent fibres that show a basal discharge are able also to convey neural information related to hypoxia intensity. Most, but not all, type I cells receive afferent innervation from one or more fibres of the carotid sinus nerve and both calyx and bouton synaptic configurations have been described. Each single fibre can oppose more than one type I cell and from more than one glomerulus. The fibre diameter ranges from 1–10 μm and the conduction velocity between 5–50 ms^{-1}. The complex innervation of the carotid body could lead to a form of elementary signal processing via post-synaptic summation or action potential collision, but there have been very few studies directed at this question and chemoreceptor discharge is most often reported simply as a mean level of frequency, although time-related information does exist, as reported *in vivo* [3].

Historically, carotid body chemoreceptor responses, recorded *in vivo*, have been correlated against arterial PO_2 levels and responses *in vitro* against the superfusate PO_2. However, non-endothelial cells sense tissue PO_2 which is

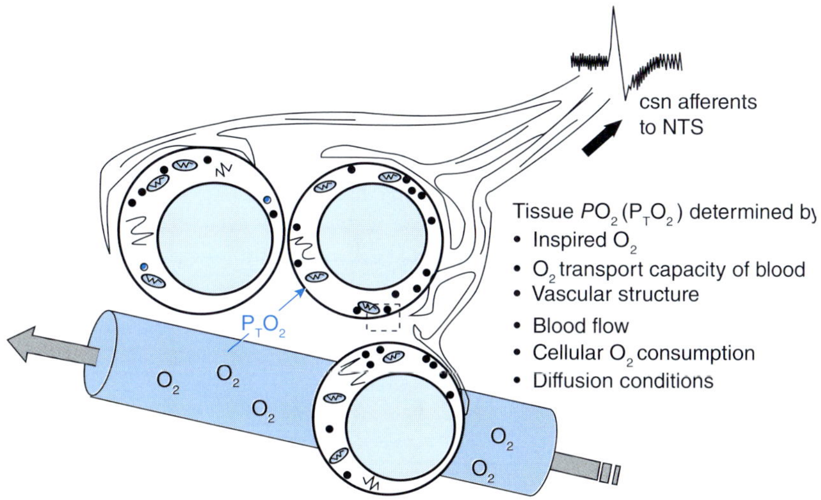

Figure 2. The functional oxygen sensing unit of the carotid body
The carotid body consists of thousands of type I cells arranged into functional clusters (glomeruli) of 3–5 cells that are able to transduce a P_TO_2 (tissue PO_2) into afferent neural discharge in the csn (carotid sinus nerve) that passes to the NTS of the medulla for the initiation of cardiorespiratory reflexes. P_TO_2 is a variable, determined by the factors listed with the 'source' of oxygen being the arterial blood and the mitochondria being the 'sink'. For clarity, type II cells have not been included in the diagram, but these are glial-like cells that sheath the type I cells and appear to offer a secondary, supportive role in oxygen sensing. Type I cells are typically around 10 μm in diameter. The small dashed rectangle indicates the area magnified in Figure 4.

lower than arterial and dependent upon the inspired PO_2, the oxygen transport capacity of the blood and the particular tissue vascular structure and blood flow, as well as, ultimately, the local oxygen consumption and diffusion conditions (Figure 2). A microvascular PO_2 of around 50 mmHg (7.5 mmHg=approx. 1 kPa) has been reported using a non-invasive phosphorescence quenching technique [2], a value that broadly agrees with earlier determinations made using sharp electrodes [4]. Thus the blood flow of the carotid body appears well matched to its metabolism and it does not appear to have a particularly low tissue PO_2 that might account for its distinct sensitivity to arterial PO_2. Instead, this suggests the existence of some inherent cellular mechanism with an apparent low affinity for oxygen. It should be remembered, however, that there exists a rich sympathetic innervation of the carotid body vasculature that might play a role in determining how this tissue PO_2 varies during hypoxia. Thus the precise horizontal position of any PO_2 stimulus-response determination will depend principally upon the point at which the PO_2 was measured. Measurements of single, type I cell acute responses to hypoxia *in vitro* would not therefore be expected to occur until the PO_2 of the medium falls below approx. 40 mmHg, with a lower value of less than approx. 10 mmHg forcing these cells into anaerobic metabolism [2] and perhaps potentially triggering at that point another set of non physiological responses. Stimulus-response curves measured *in vivo*, with arterial blood PO_2 taken as the stimulus, would thus be right-shifted relative to single cell, *in vitro* responses (Figure 3).

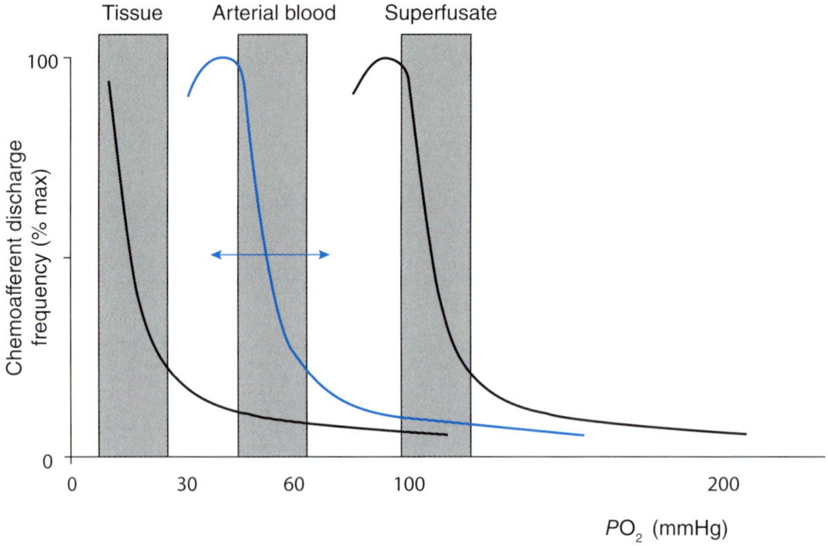

Figure 3. The horizontal position of the PO_2 stimulus-response curve is determined by the point of measurement of PO_2

The chemoafferent neural response to decreasing PO_2 is exponential, with discharge only failing at severe levels of hypoxia. Three identical responses are shown. The precise position of the curve along the PO_2 axis will be determined by the measurement of PO_2 and will left-shift as this measurement becomes more closely related to tissue PO_2. Thus in many *in vitro* organ preparations the PO_2 of the superfusate grossly overestimates the tissue PO_2 within the core of the organ (curve to far right). The overestimation will be determined by the depth from the organ surface of the cell recorded from and the local diffusion/metabolism conditions. A general rule of thumb for a highly metabolic tissue such as the carotid body, would be approx. 1 mmHg fall in PO_2 per 1 μm depth from the surface. Although the overestimation is reduced *in vivo*, arterial blood sampling still produces a curve that is right-shifted (centre curve) relative to the precise stimulus being sensed. The curve at the far left, indicates a theoretical curve were P_TO_2 able to be recorded simultaneously with afferent discharge. Horizontal shifts have been approximated from literature values. The vertical bars indicate the area of greatest sensitivity to PO_2 and, although not actually different in all three cases, do appear different as a consequence of the point of PO_2 measurement. The horizontal arrow on the middle curve indicates the range of sensitivities possible for different afferent neurones of the carotid sinus nerve.

In addition to reductions in the arterial PO_2 (i.e. hypoxic hypoxia), stagnant or ischaemic hypoxia and histotoxic hypoxia are also powerful stimuli of the carotid body. Thus sustained and significant haemorrhage leading to hypotension and metabolic poisons including cyanide all increase chemoafferent discharge. In contrast, the arterial content of oxygen does not appear to be sensed by the carotid body of all species and anaemic hypoxia, the reduction of blood oxygen content, without alteration in arterial PO_2, is not a principal stimulus of the carotid body and therefore does not stimulate ventilation. Accordingly, carbon monoxide at non-lethal arterial tensions also does not stimulate ventilation and carbon monoxide applied to the carotid body may even, through elevating arterial PO_2, decrease chemoafferent discharge [5]. The degree of binding of carbon monoxide to haemoglobin (termed HbCO) needs to be >20% before an increase in discharge is noted but even up to 50%

HbCO, the effect is small and an explanation for this lack of excitation by HbCO may lie with the relatively high blood flow of the carotid body and thus its intrinsic inability to sense oxygen delivery itself. The rodent carotid body may differ in this respect, with one report [5a] suggesting that the rat carotid body might detect the arterial content of oxygen; although, to date, this has only been reported at the level of ventilation. The mechanism underlying this species difference is not clear.

Chemoafferent and reflex responses to hypoxia

Chemoafferent discharge provides the final common output of the carotid body and could be considered therefore as a definitive measure of its response to hypoxia. A tonic activity of around 1–2 Hz is observed in carotid sinus nerve single-fibre chemoafferents during hyperoxia *in vivo* with no absolute physiological threshold to arterial PO_2. The discharge is aperiodic with interspike interval distributions fitting a Poisson distribution. The response to hypoxic hypoxia is rapid with a latency of <1 s and a time to peak of 1–5 s, after subtraction of circulatory transit delays. This response does not adapt and remains relatively sustained for the duration of the hypoxia episode, provided the intensity is not too severe. Steady-state discharge rises increasingly with mounting hypoxia, at a constant arterial PCO_2 (partial pressure of carbon dioxide), and may be best described by single exponential function with an offset to account for the lack of an oxygen threshold. Chemoreceptor discharge begins to rise gradually from between 400–140 mmHg arterial PO_2 until approx. 100 mmHg when the slope increases and then rises more abruptly at an arterial PO_2 of approx. 70–75 mmHg, reaching a maximum activity at an arterial PO_2 of approx. 20–30 mmHg. A reduction in discharge frequency at an arterial PO_2 of below approx. 20 mmHg has also been observed and thus the full arterial PO_2-response curve may therefore best be described by a sigmoidal function, which is also positioned appropriately as an inverse function of the haemoglobin oxygen-dissociation curve.

Single-fibre afferent frequency is elevated by 10–20 times basal at its peak across most species studies, with the *in vivo* absolute level appearing species-dependent and difficult to quantify precisely due to various degrees of anaesthesia/paralytic agents. Different fibres appear to have different thresholds and peaks and the inflection in the curves of different chemoreceptor fibres can occur over the arterial PO_2 range of 60–110 mmHg [6], reflecting perhaps variability in the affinity range of an oxygen sensor(s) or to gradients in tissue PO_2 within the organ. At present, there is no compelling evidence to suggest that a single oxygen sensor is responsible for the entire stimulus-response characteristics of the carotid body and a compelling alternative is that a range of sensors with various affinities and thresholds might better explain the response [7] or, indeed, provide an important failsafe redundancy. Recordings from multi-fibre rather than single-fibre preparations are therefore prone to errors if used for determining sensitivity at a single-cell level and should only be interpreted with caution to infer transduction processes.

In the adult, the reflex response to hypoxia includes hyperventilation, i.e. an increase in alveolar ventilation above metabolic demand, which acts to decrease alveolar PCO_2 levels and thus increase alveolar (and hence arterial) PO_2 levels towards that of inspired. In addition to respiratory reflexes, an elevated chemoafferent discharge can also induce direct cardiac and vascular reflex responses that include bradycardia, decreased cardiac output and peripheral vasoconstriction as well as augmented adrenomedullary output, but these effects are often masked by secondary feedback responses, via thoracic afferents and baroreceptors, subsequent to the hyperventilation [8]. These complex reflexes and interactions hint at how carotid body dysfunction might be implicated in the aetiology of a variety of cardiorespiratory disorders, including sleep apnoea, hypertension and cardiac arrhythmia although a definitive cause-effect has not been established.

Calcium-dependent neurotransmission

Hypoxia induces intracellular calcium elevations in isolated type I cells and neurosecretion from these cells is calcium-dependent. Although a role for intracellular calcium stores is not entirely without experimental support, the elevated intracellular calcium during hypoxia arises predominantly subsequent to calcium entry via membrane channels [9]. Thus, in the absence of external calcium, chemoafferent discharge responses to hypoxia are greatly attenuated. Attenuation in discharge [10] and in catecholamine secretion [11] is also observed, and in a dose-dependent manner, with increasing application of a variety of calcium channel blockers, including nifedipine, suggesting an important role for entry via L-type VGCCs (voltage-gated calcium channels). This common finding implicates cell depolarization as an essential requirement of hypoxia sensing. Additionally, and crucially, mitochondrial membrane depolarization during hypoxia occurs after cell membrane depolarization [12]. Finally, confirming a role for VGCCs, when the type I cell membrane potential is clamped at its resting membrane potential, the elevation in calcium during hypoxia/anoxia is attenuated and slowed [9]. The calcium response is graded with hypoxia intensity but it is worth noting that the relationship between calcium and neurosecretion need not be linear. In addition to L-type, the presence of N-, P-/Q- and R-type calcium channels have also been demonstrated in chemoreceptor cells with, in the rabbit, perhaps only L- and P-/Q-activation being coupled to neurosecretion [13]. More recently, TRP (transient receptor potential) channels have been localized to the carotid body [14] and a voltage-independent entry pathway for calcium during hypoxia may therefore also exist. More work is required to establish the role of these channels in oxygen sensing.

Carotid body type I cells contain a variety of both excitatory and inhibitory neurotransmitter substances. These range from amines including catecholamines and acetylcholine to neuropeptides, including substance P, the purines ATP and adenosine, amino acids, including GABA (γ-aminobutyric acid) and even the gas signalling molecules, nitric oxide and carbon monoxide. A detailed description of the effects of these various substances on carotid body function is not possible here and can be found elsewhere [15], but it is becoming apparent that a single transmitter cannot account for the hypoxia response of the carotid body [7]. In addition, a considerable species differences is apparent in the specific transmitter used to signal hypoxia and different species may even use the same transmitter in dissimilar ways. Thus dopamine and acetylcholine have been shown to have different effects upon excitation in cats and rabbits. The consequences of these differences, if there are any, are not known. The presence of pre-synaptic (i.e. type I cell) autoreceptors, in addition to post-synaptic receptors, reveals a complexity of control that is not yet fully understood. For example, adenosine, released from type I cells during hypoxia, can act on type I cells to inhibit a 4-aminopyridine-sensitive outward potassium current [16]. The role of this in shaping the response to hypoxia is not known, but it presumably acts to augment the response.

Of the various transmitters postulated to be central to the transduction process, attention has focussed in recent years upon the co-transmitter substance, ATP. This follows the finding that ATP-activated purinoceptors, specifically the ligand-gated P2X ion channel subtype, exists on chemoafferent nerve terminals and type I cells [17], that blocking the purinergic receptor attenuates (and inhibits when combined with blockage of the cholinergic receptor) the sensory response to hypoxia [18] and that modified mice lacking the $P2X_2$, but not $P2X_3$ receptors, have a blunted hypoxia sensitivity at both the ventilatory and chemoafferrent level [19]. P2X receptors desensitize rapidly and although this may be an important information-containing process in central neurones, it does not appear consistent with the sustained chemoafferent discharge observed during a period of hypoxia. Thus interplay between both excitatory and inhibitory transmitters might exist to prolong the excitation and Prabhakar [7] has forwarded what he calls a 'push-pull' hypothesis to account for this phenomenon. This interesting hypothesis has yet to be tested. Additionally, the existence of various transmitter substances might be an essential requirement for the polymodal receptor nature of the carotid body type I cell [20] as well as for enabling plasticity in response to changes during development or in its tissue environment due, for example, to chronic diseases predisposing to alterations in blood-gas chemistry. The precise role of each of the chemical substances in the carotid body in either determining post-synaptic action potential frequency or in shaping the response is not yet known with certainty but should prove a demanding challenge in the future if we are to understand fully how the carotid body functions.

Potassium channel inhibition and depolarization

The absolute requirement for VGCCs implicates membrane depolarization as a key upstream event in the hypoxia transduction process and this is believed to be achieved by closure of potassium channels located in the plasma membrane of type I cells. These channels must therefore have the property of being either directly, or indirectly, inactivated by hypoxia and must have a finite open conductance at the E_m (resting membrane potential; RMP), which, in the rat type I cell, is between −45 and −55 mV. The first description of a fast inactivating, delayed-rectifier potassium current in rabbit type I cells that was decreased, reversibly, by hypoxia was made almost 20 years ago [21]. This effect appeared to be membrane delimited, suggesting an intrinsic sensor mechanism that might therefore be associated with either the α-pore forming subunits of the channel or with associated, auxiliary β-subunits. Later attempts at recreating oxygen sensitivity using recombinant α- and β-subunits in cell expression systems did not, however, prove conclusive and, together with studies in other species demonstrating potassium channel oxygen sensitivity only in patch clamp configurations that would retain cytosolic or closely-associated factors, have returned the focus to identifying an alternative sensor mechanism that couples to potassium channels.

Although a number of potassium channel subtypes are now known to be sensitive to hypoxia, there appears to be a significant species difference, as well as discrepancies within a species, and a definitive identity of the channel(s) responsible for initiating depolarization in type I cells has yet to be made. There is, however, no absolute requirement for just one class of potassium channel to be involved in oxygen sensing and all types may play a role within a single cell to both induce and maintain the depolarization. Thus, in addition to the originally-described Kv (voltage-dependent potassium channels; most likely the Kv4 subtype in rabbit and the Kv3 subtype in mice [22]), non-inactivating, calcium-dependent potassium channels (principally the large conductance, BK subtype [23]), hERG potassium channels [24] and a heteromeric combination of TASK1 and 3 subunits of the voltage-independent, acid-sensitive tandem pore domain potassium channel family [25] have also been forwarded as candidate oxygen-dependent ion channels. Of these, PO_2-response curves for channel inactivation that correlate with the cellular response to tissue hypoxia have been demonstrated convincingly only with the TASK channels that have a $K_{1/2}$ of 12–13 mmHg PO_2. These channels have a low (sub 20 pS) conductance and are activated at potentials greater than approx. −90 mV, making them suitable for the initiation of hypoxia-mediated depolarization. Selective pharmacological approaches to studying these channels are limited at present, with barium, quinidine and both local and gaseous anaesthetics being the prime agents and, more recently, the cannabinoid, anandamide. The gating of these channels by other aspects of cellular metabolism, including pH and lipids, add complexity to the interpretation, but it is noteworthy that barium can induce carotid body afferent discharge [26]. Conversely, the selective blockers of BK channels, namely iberiotoxin and charybdotoxin, are without effect upon resting carotid body chemoafferent discharge or membrane potential [26] as would have been expected were they contributing to the RMP. Additionally, the relatively non-selective, but TASK-independent, potassium channel blockers TEA (tetraethylammonium) and 4-AP (4-aminopyridine), do not appear to induce afferent discharge or prevent hypoxia-induced discharge and it is therefore not easy to reconcile these findings with others that show a depolarization in type I cells during normoxic, current clamp with charybdotoxin [27]. Some of the discrepancy might exist due to different preparations and this explanation appears founded as a depolarizing and neurosecretory effect of TEA in a novel slice preparation of the carotid body in which cell–cell interactions would be more likely preserved than in the previous isolated cell experiments has been described [28]. It has also been reported that, although charybdotoxin was without effect upon resting chemoafferent discharge, in the presence of hypoxia, this blocker of BK channels right-shifted the PO_2-response curve, implying a greater effect with increasing membrane depolarization [29]. Thus were the large conductance BK channels not inhibited by hypoxia, the small, depolarizing currents of approx. 7 pA induced by TASK channel closures, would be rapidly reversed, potentially preventing a sustained calcium influx and perhaps leading to an adapting hypoxia response, which is not observed *in vivo*. On the other hand, the process of neurosecretion does appear to initiate a limiting control on TASK channels as it has been shown that $GABA_B$ autoreceptors can activate TASK-like channels on type I cells thus potentially reversing hypoxia-induced depolarization [30]. It is perhaps worth noting that up to 10% of type I cells might not respond to hypoxia and in those that do the magnitude of depolarization in response to hypoxia is generally quite small, at around 10 mV. The reason, or function, of this heterologous response is not apparent, nor why, within and across species, variations exist in oxygen sensitive potassium channels and the case for an essential requirement for potassium channel inhibition in hypoxia transduction, whilst compelling and widely accepted, is not yet fully proven.

The nature of the oxygen sensor

The sigmoidal shape of the PO_2 afferent nerve response curve and its similarity to the haemoglobin-oxygen dissociation curve suggests that the sensor of the lack of oxygen might be a haem protein, an idea borne out by the finding that most hypotheses of hypoxia transduction invoke the participation of such proteins, from either intra- or extra-mitochondrial locations. The relative importance of these various proteins to the full functional response of the carotid body would ultimately depend upon their oxygen affinity and the nature of their coupling to the transduction process and it is important to re-iterate that the role of any putative sensors need not be mutually exclusive.

The oxygen affinity of mitochondrial proteins of the electron transport chain in general appears too high for a role in sensing systemic levels of hypoxia, yet a role for these complexes has long been suggested based, principally, upon the finding that blockers of the electron transport chain or uncouplers of oxidative phosphorylation are potent chemostimuli, increasing afferent chemodischarge and intracellular calcium and inhibiting TASK channels of type I cells [31]. Interestingly, the inhibition of TASK channels by hypoxia appears cell specific [32] and thus type I cell mitochondria may possess some, as yet undefined, specialization, as suggested by the particularly exquisite sensitivity to hypoxia of their membrane potential and NAD(P)H autofluorescence [33] and/or the activity of cytochrome *c* oxidase in complex IV of the electron transport chain [34]. Importantly for this hypothesis, hypoxia responsiveness is greatly reduced or abolished in the presence of the various blockers and uncouplers, which in itself, suggests some commonality of mechanism, although care needs to be exercised when interpreting some of these data, especially when the complex I blocker, rotenone, has been used as this appears to have undefined extra-mitochondrial effects upon oxygen sensing [35].

If the mitochondria are acting as oxygen sensors, the question arises as to how they couple their reduced activity during hypoxia to the plasma membrane. Current mechanisms proposed include variations in the cellular redox potential, consequent to variations in the concentration of ROS (reactive oxygen species), including the superoxide anion, O_2^-, formed by incomplete electron transfer in the electron transport chain (principally in complexes I and III) and its diffusible, dismutased product, hydrogen peroxide, H_2O_2. Although ROS may affect potassium channel function, their role in oxygen sensing is contentious [36] with global cellular redox apparently unchanged during hypoxia. Another source of ROS localized more closely to plasmalemmal potassium channels is the extra-mitochondrial, membrane-bound NADPH oxidase complex, which produces O_2^- and H_2O_2 in proportion to PO_2 [37]. Although proposed as a sensor mechanism, an action of NADPH oxidase-derived ROS on type I cell potassium channel inactivation is not proven. Thus, whilst subunits of NADPH oxidase are located within type I cells, mice with a selective (gp91^{phox}) gene deletion of NADPH oxidase activity retained normal type I cell and carotid body function in response to hypoxia [38,39]. Further dispelling a role for NADPH oxidase in sensing hypoxia, is the more recent finding that targeted deletion of the p47^{phox} subunit of NADPH oxidase led to an enhanced inactivation of type I cell BK channels [40] and this suggests that ROS may be involved in cell repolarization during the recovery from hypoxia, rather than depolarization. Although an action on other potassium channels in the carotid body has not been confirmed to date, a role for ROS in mediating hypoxia chemotransduction is not compelling.

An alternative signalling pathway to couple mitochondrial function with potassium channel inhibition may be via changes in the cytosolic concentration of ATP or factor(s) related to ATP synthesis. This is a long established hypothesis and is based upon a predicted final common consequence of hypoxia and electron transport chain blocking and uncoupling drugs and includes evidence demonstrating an excitatory action upon chemodischarge of the mitochondrial ATPase inhibitor, oligomycin, [41] and more recently and specifically, inhibitory actions of carbon monoxide, rotenone, myxothiazol, cyanide, DNP (2,4-dinitrophenol) and FCCP (carbonyl-p-trifluoromethoxyphenylhydrazone) upon TASK channel activation in type I cells [31,42]. The common feature of all of these agents would be a predicted fall in cytosolic ATP concentration. These findings, additionally, argue against an important role for ROS as these levels would be affected differently by these various agents. A role for mitochondria in oxygen sensing is, however, not without criticism and it has been reported that, other than rotenone, many inhibitors of oxidative phosphorylation have non-specific actions upon potassium channel activity that are independent of the effects of hypoxia [35]. In addition, partial deletion of *sdhd*, a component of mitochondrial complex II, in mice was without effect upon hypoxia sensitivity as assayed by catecholamine release from type I cells [43]. Reconciling these differences will require further experimentation and it remains to be seen whether type I cell mitochondria are specialized for oxygen sensing in the carotid body.

Sensing systemic hypoxia is an energy consuming process and, not surprisingly, cellular ATP levels do not fall until the cell is severely compromised. Thus, although a direct action of ATP upon TASK channels has been described in isolated membrane patches [44], it seems likely that this occurs indirectly in the whole cell. One such mechanism is consequent to the activity of adenylate kinase which, by converting two molecules of ADP into ATP and AMP, increases the cellular AMP:ATP ratio, raising AMP levels sufficiently to stimulate *de novo* synthesis of AMPK (AMP-activated protein kinase). AMPK is ubiquitously expressed in eukaryotic cells and plays a vital role in cellular metabolic regulation. Cellular stressors, including hypoxia, which decrease the energy status of the cell [45] activate AMPK which functions to augment ATP-generating processes while inhibiting non-essential energy consuming processes. Recently, a novel downstream target for this kinase, namely potassium channels of the carotid body type I cell, was identified and a hypothesis that places AMPK activation at the heart of oxygen sensing has been proposed [46]. Thus, AMPK, located within approx. 1 μm of the plasma membrane in type I cells, can, subsequent to its [AMP]-independent activation by the drug, AICA (5-aminoimidazole-4-carboxamide) riboside, reversibly cause both BK and TASK channel inactivation, VGCC-dependent elevations in intracellular calcium and increased afferent neural discharge [47]. These effects could also all be reversed or inhibited by the newly synthesized, AMPK antagonist, compound C. This hypothesis is particularly attractive as it unites the mitochondrial and potassium channel responses to hypoxia without a need to necessarily invoke mitochondrial specialization or decreases in [ATP], neither of which has compelling evidence for its support.

An implication, however, if one assumes that the type I cell has a sensitivity to hypoxia not observed in other non-oxygen sensing tissues, must be that the carotid body either possesses unique isoform complexes of AMPK or that signalling by AMPK is somehow modified for a specific function. Further experiments will be required to ascertain the basis for this organ's unique oxygen-sensing ability.

In contrast to the hypotheses requiring alterations in the cellular redox or energy status, another proposal has been forwarded that assumes a key role for gas molecule signalling in the inactivation of BK channels. Thus, Kemp and co-workers [48] were able to demonstrate that the oxygen-dependent enzyme, HO-2 (haemoxygenase-2) was closely associated in a complex with membrane-bound potassium channels and that at least one of its products, carbon monoxide, appeared essential for BK channel activation. According to this model, in the presence of oxygen and its co-substrates, haem and NADPH, carbon monoxide increases BK channel open probability, but when oxygen levels are compromised the reduction in carbon monoxide production would lead to BK inactivation and membrane depolarization. Carbon monoxide is a remarkably inert molecule, interacting almost exclusively with haem groups, of which the haem-binding domain of the α-subunit of BK appears the most likely site of its action. These data support earlier reports of an inhibitory function of HO-generated carbon monoxide at the carotid body [49,50] and further experiments are keenly anticipated.

Conclusion

The critical importance of matching oxygen delivery to metabolic rate is dependent upon a controlled reflex response to hypoxia and an understanding of this process may go some way towards a more full appreciation of the mechanisms and consequences of cardiorespiratory disease states. A full understanding of carotid body chemotransduction processes would be an important contribution to this appreciation and appears tantalisingly close. Consensus exists on the importance of calcium-dependent neurosecretion with the elevated intracellular calcium arising via VGCCs activated by membrane depolarization, induced by potassium channel closure subsequent to the detection of reduced oxygen by an intracellular protein sensor. The major focus of interest presently is in defining the oxygen sensor and several mitochondrial and non-mitochondrial candidate proteins have been proposed of which, presently, the cytosolic AMPK and the plasma membrane-bound HO-2 are prime candidates (Figure 4). These may all be important in shaping the full response to various degrees of hypoxia and it is hoped that future studies might attempt to begin a reconciliation of the various hypotheses. One way to aid this would be in the generation of full response curves for each putative sensor, from which oxygen affinities could be better determined.

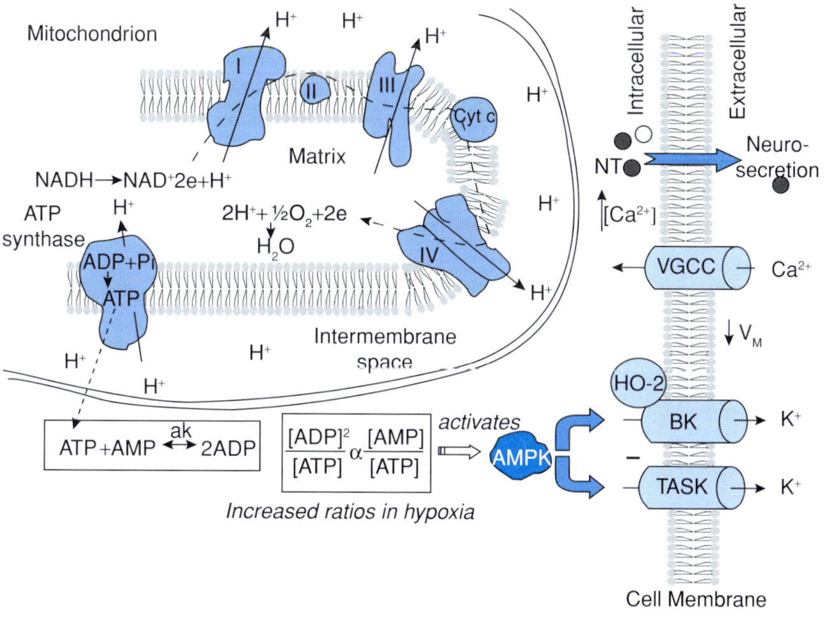

Figure 4. Coupling the mitochondrial and membrane hypotheses of chemotransduction

In oxidative phosphorylation, mitochondrial respiratory complexes (I–IV) shuttle electrons (dashed line) to molecular oxygen, which is reduced to water. The proton gradient produced by this process is used by ATP synthase to generate ATP. Adenylate kinase (ak) catalyses the conversion of 2ADP into ATP and AMP and the ratio of ADP/ATP is thus the square of the AMP/ATP ratio. Increases in the AMP/ATP ratio are therefore sensitive to changes in the 'energy status' of the cell, even in the absence of measurable changes in ATP concentration. The AMP/ATP ratio increases during hypoxia and activates AMPK that acts to inhibit cellular ATP-consuming catabolic processes and to augment ATP-producing anabolic processes. In addition, AMPK can inactivate potassium channels of both the TASK and the BK subtypes that leads to membrane depolarization (decreased V_M). The magnitude of this depolarization is sufficient to activate, primarily, L-type VGCCs that raises intracellular calcium concentration and induces neurotransmitter (NT) release on to post-synaptic afferent nerve endings (not shown). Additionally, carbon monoxide production by HO-2 is decreased or inhibited during hypoxia, thus removing its activating influence upon K_{Ca}, facilitating the depolarization.

Summary

- *The carotid body is a peripheral oxygen sensor located in the bifurcation of the internal and external carotid arteries that senses arterial hypoxia translating this into afferent neural discharge that initiates corrective cardiorespiratory reflexes.*
- *Afferent discharge occurs subsequent to calcium dependent neurosecretion from type I cells on to closely apposed free nerve endings.*
- *The elevation in intracellular calcium is dependent upon calcium entry into type I cells through VGCCs and this is consequent to cell depolarization following the hypoxia-inactivation of specific, species-dependent classes of potassium channels, including TASK and BK.*

- *The sensitivity of the carotid body to hypoxia appears high and suggests the presence of a low-affinity oxygen sensor system(s).*
- *The nature of the oxygen sensor is not known for certain. The generation or reduction of ROS during hypoxia appears not to be a vital part of the transduction process but the proteins AMPK and HO-2 both have been put forward as essential components of this pathway and both have been associated with potassium channel inactivation during hypoxia.*

I thank the Lister Institute, the Wellcome Trust and the British Heart Foundation for their generous support of my research and my colleagues for their collaboration over the years. Special thanks also go to Dr Joanne Wilton, University of Cambridge, for providing Figure 1.

References

1. McDonald, D.M. (1981) Peripheral chemoreceptors. Structure-function relationships of the carotid body, in: *Regulation of Breathing, Volume 17, Lung Biology in Health and Disease* (Hornbein, T.F., ed.), pp. 105–319, Dekker, New York
2. Rumsey, W.L., Iturriaga, R., Spergel, D., Lahiri S. & Wilson, D.F. (1991) Optical measurements of the dependence of chemoreception on oxygen pressure in the cat carotid body. *Am. J. Physiol.* **261**, C614–C622
3. Kumar, P., Nye, P.C. & Torrance, R.W. (1988) Do oxygen tension variations contribute to the respiratory oscillations of chemoreceptor discharge in the cat? *J. Physiol.* **395**, 531–552
4. Whalen, W.J., Savoca, J. & Nair, P. (1973) Oxygen tension measurements in carotid body of the cat. *Am. J. Physiol.* **225**, 986–991
5. Nye, P.C., Torrance, R.W. & Folgering, H. (1982) Arterial chemoreceptor inhibition by a single inspirate containing carbon monoxide is accounted for by raised arterial PO_2. *Pflugers Arch.* **393**, 313–317

5a. Garland, R.J., Kinkead, R. & Milsom, W.K. (1994) The ventilatory response to changes in arterial oxygen content. *Respir. Physiol.* **96**, 199–211

6. Vidruk, E.H., Olson, Jr, E.B., Ling, L. & Mitchell, G.S. (2001) Responses of single-unit carotid body chemoreceptors in adult rats. *J. Physiol.* **531**, 165–170
7. Prabhakar, N.R. (2006) O_2 sensing at the mammalian carotid body: why multiple O_2 sensors and multiple transmitters? *Exp. Physiol.* **91**, 17–23
8. Kara, T., Narkiewicz, K. & Somers, V.K. (2003) Chemoreflexes-physiology and clinical implications. *Acta Physiol. Scand.* **177**, 377–384
9. Buckler, K.J. & Vaughan-Jones, R.D. (1994) Effects of hypoxia on membrane potential and intracellular calcium in rat neonatal carotid body type I cells. *J. Physiol.* **476**, 423–428
10. Shirahata, M. & Fitzgerald, R.S. (1991) Dependency of hypoxic chemotransduction in cat carotid body on voltage-gated calcium channels. *J. Appl. Physiol.* **71**, 1062–1069
11. Obeso, A., Rocher, A., Fidone, S. & Gonzalez, C. (1992) The role of dihydropyridine-sensitive Ca^{2+} channels in stimulus-evoked catecholamine release from chemoreceptor cells of the carotid body. *Neuroscience* **47**, 463–472
12. Buckler, K.J. & Vaughan-Jones, R.D. (1998) Effects of mitochondrial uncouplers on intracellular calcium, pH and membrane potential in rat carotid body type I cells. *J. Physiol.* **513**, 819–833
13. Rocher, A., Geijo-Barrientos, E., Caceres, A.I., Rigual, R., Gonzalez, C. & Almaraz, L. (2005) Role of voltage-dependent calcium channels in stimulus-secretion coupling in rabbit carotid body chemoreceptor cells. *J. Physiol.* **562**, 407–420
14. Buniel, M.C., Schilling, W.P. & Kunze, D.L. (2003) Distribution of transient receptor potential channels in the rat carotid chemosensory pathway. *J. Comp. Neurol.* **464**, 404–413

15. Lahiri, S., Roy, A., Baby, S.M., Hoshi, T., Semenza, G.L. & Prabhakar, N.R. (2006) Oxygen sensing in the body. *Prog. Biophys. Mol. Biol.* **91**, 249–286
16. Vandier, C., Conway, A.F., Landauer, R.C. & Kumar, P. (1999) Presynaptic action of adenosine on a 4-aminopyridine-sensitive current in the rat carotid body. *J. Physiol.* **515**, 419–429
17. Prasad, M., Fearon, I.M., Zhang, M., Laing, M., Vollmer, C. & Nurse, C.A. (2001) Expression of P2×2 and P2×3 receptor subunits in rat carotid body afferent neurones: role in chemosensory signalling. *J. Physiol.* **537**, 667–677
18. Zhang, M., Zhong, H., Vollmer, C. & Nurse, C.A. (2000) Co-release of ATP and ACh mediates hypoxic signalling at rat carotid body chemoreceptors. *J. Physiol.* **525**, 143–158
19. Rong, W., Gourine, A.V., Cockayne, D.A., Xiang, Z., Ford, A.P., Spyer, K.M. & Burnstock, G. (2003) Pivotal role of nucleotide P2×2 receptor subunit of the ATP-gated ion channel mediating ventilatory responses to hypoxia. *J. Neurosci.* **23**, 11315–11321
20. Kumar, P. & Bin-Jaliah, I. (2007) Adequate stimuli of the carotid body: more than an oxygen sensor? *Respir. Physiol. Neurobiol.* **157**, 12–21
21. Lopez-Barneo, J., Lopez-Lopez, J.R., Urena, J. & Gonzalez, C. (1988) Chemotransduction in the carotid body: K^+ current modulated by Po_2 in type I chemoreceptor cells. *Science* **241**, 580–582
22. Perez-Garcia, M.T., Colinas, O., Miguel-Velado, E., Moreno-Dominguez, A. & Lopez-Lopez, J.R. (2004) Characterization of the Kv channels of mouse carotid body chemoreceptor cells and their role in oxygen sensing. *J. Physiol.* **557**, 457–471
23. Peers, C. (1990) Hypoxic suppression of K^+ currents in type I carotid body cells: selective effect on the Ca^{2+}-activated K^+ current. *Neurosci. Lett.* **119**, 253–256
24. Overholt, J.L., Ficker, E., Yang, T., Shams, H., Bright, G.R. & Prabhakar, N.R. (2000) HERG-like potassium current regulates the resting membrane potential in glomus cells of the rabbit carotid body. *J. Neurophysiol.* **83**, 1150–1157
25. Buckler, K.J., Williams, B.A. & Honore, E. (2000) An oxygen-, acid- and anaesthetic-sensitive TASK-like background potassium channel in rat arterial chemoreceptor cells. *J. Physiol.* **525**, 135–142
26. Donnelly, D.F. (1997) Are oxygen-dependent K^+ channels essential for carotid body chemo-transduction? *Respir. Physiol.* **110**, 211–218
27. Wyatt, C.N. & Peers, C. (1995) Ca^{2+}-activated K^+ channels in isolated type I cells of the neonatal rat carotid body. *J. Physiol.* **483**, 559–565
28. Pardal, R., Ludewig, U., Garcia-Hirschfeld, J. & Lopez-Barneo, J. (2000) Secretory responses of intact glomus cells in thin slices of rat carotid body to hypoxia and tetraethylammonium. *Proc. Natl. Acad. Sci. U.S.A.* **97**, 2361–2366
29. Pepper, D.R., Landauer, R.C. & Kumar, P. (1995) Effect of charybdotoxin on hypoxic chemosensitivity in the adult rat carotid body *in vitro*. *J. Physiol.* **487P**, 177P
30. Fearon, I.M., Zhang, M., Vollmer, C. & Nurse, C.A. (2003) GABA mediates autoreceptor feedback inhibition in the rat carotid body via presynaptic $GABA_B$ receptors and TASK-1. *J. Physiol.* **553**, 83–94
31. Wyatt, C.N. & Buckler, K.J. (2004) The effect of mitochondrial inhibitors on membrane currents in isolated neonatal rat carotid body type I cells. *J. Physiol.* **556**, 175–191
32. Johnson, R.P., O'Kelly, I.M. & Fearon, I.M. (2004) System-specific O_2 sensitivity of the tandem pore domain K^+ channel TASK-1. *Am. J. Physiol. Cell. Physiol.* **286**, C391–C397
33. Biscoe, T.J. & Duchen, M.R. (1990) Responses of type I cells dissociated from the rabbit carotid body to hypoxia. *J. Physiol.* **428**, 39–59
34. Wilson, D.F., Mokashi, A., Chugh, D., Vinogradov, S., Osanai, S. & Lahiri, S. (1994) The primary oxygen sensor of the cat carotid body is cytochrome a3 of the mitochondrial respiratory chain. *FEBS Lett.* **351**, 370–374
35. Ortega-Saenz, P., Pardal, R., Garcia-Fernandez, M. & Lopez-Barneo, J. (2003) Rotenone selectively occludes sensitivity to hypoxia in rat carotid body glomus cells. *J. Physiol.* **548**, 789–800
36. Gonzalez, C., Sanz-Alfayate, G., Agapito, M.T., Gomez Nino, A., Rocher, A. & Obeso, A. (2002) Significance of ROS in oxygen sensing in cell systems with sensitivity to physiological hypoxia. *Respir. Physiol. Neurobiol.* **132**, 17–41

37. Acker, H. (1994) Mechanisms and meaning of cellular oxygen sensing in the organism. *Respir. Physiol.* **95**, 1–10
38. Roy, A., Rozanov, C., Mokashi, A., Daudu, P., Al-mehdi, A.B., Shams, H. & Lahiri, S. (2000) Mice lacking in gp91 phox subunit of NAD(P)H oxidase showed glomus cell $[Ca^{2+}]_i$ and respiratory responses to hypoxia. *Brain Res.* **872**, 188–193
39. He, L., Chen, J., Dinger, B., Sanders, K., Sundar, K., Hoidal, J. & Fidone, S. (2002) Characteristics of carotid body chemosensitivity in NADPH oxidase-deficient mice. *Am. J. Physiol. Cell Physiol.* **282**, C27–C33
40. He, L., Dinger, B., Sanders, K., Hoidal, J., Obeso, A., Stensaas, L., Fidone, S. & Gonzalez, C. (2005) Effect of p47phox gene deletion on ROS production and oxygen sensing in mouse carotid body chemoreceptor cells. *Am. J. Physiol. Lung Cell Mol. Physiol.* **289**, L916–L924
41. Mulligan, E. & Lahiri, S. (1981) Dependence of carotid chemoreceptor stimulation by metabolic agents on PaO_2 and $PaCO_2$. *J. Appl. Physiol.* **50**, 884–891
42. Barbe, C., Al-Hashem, F., Conway, A.F., Dubuis, E., Vandier, C. & Kumar, P. (2002) A possible dual site of action for carbon monoxide-mediated chemoexcitation in the rat carotid body. *J. Physiol.* **543**, 933–945
43. Piruat, J.I., Pintado, C.O., Ortega-Saenz, P., Roche, M. & Lopez-Barneo, J. (2004) The mitochondrial SDHD gene is required for early embryogenesis, and its partial deficiency results in persistent carotid body glomus cell activation with full responsiveness to hypoxia. *Mol. Cell. Biol.* **24**, 10933–10940
44. Williams, B.A. & Buckler, K.J. (2004) Biophysical properties and metabolic regulation of a TASK-like potassium channel in rat carotid body type I cells. *Am. J. Physiol. Lung Cell Mol. Physiol.* **286**, L221–L230
45. Hardie, D.G., Hawley, S.A. & Scott, J.W. (2006) AMP-activated protein kinase-development of the energy sensor concept. *J. Physiol.* **574**, 7–15
46. Evans, A.M., Mustard, K.J., Wyatt, C.N., Peers, C., Dipp, M., Kumar, P., Kinnear, N.P. & Hardie, D.G. (2005) Does AMP-activated protein kinase couple inhibition of mitochondrial oxidative phosphorylation by hypoxia to calcium signaling in O_2-sensing cells? *J. Biol. Chem.* **280**, 41504–41511
47. Wyatt, C.N., Kumar, P., Aley, P., Peers, C., Hardie, D.G. & Evans, A.M. (2006) Does AMP-activated protein kinase couple hypoxic inhibition of oxidative phosphorylation to carotid body excitation? *Adv. Exp. Med. Biol.* **580**, 191–196; discussion 351–359
48. Williams, S.E., Wootton, P., Mason, H.S., Bould, J., Iles, D.E., Riccardi, D., Peers, C. & Kemp, P.J. (2004) Hemoxygenase-2 is an oxygen sensor for a calcium-sensitive potassium channel. *Science* **306**, 2093–2097
49. Prabhakar, N.R. (1998) Endogenous carbon monoxide in control of respiration. *Respir. Physiol.* **114**, 57–64
50. Riesco-Fagundo, A.M., Perez-Garcia, M.T., Gonzalez, C. & Lopez-Lopez, J.R. (2001) O_2 modulates large-conductance Ca^{2+}-dependent K^+ channels of rat chemoreceptor cells by a membrane-restricted and CO-sensitive mechanism. *Circ. Res.* **89**, 430–436

5

Hypoxic pulmonary vasoconstriction

A. Mark Evans[1]

School of Biology, Bute Building, University of St Andrews, St Andrews, Fife KY16 9TS, U.K.

Abstract

HPV (hypoxic pulmonary vasoconstriction) is the critical and distinguishing characteristic of the arteries that feed the lung. In marked contrast, systemic arteries dilate in response to hypoxia to meet the metabolic demands of the tissues they supply. Physiologically, HPV contributes to ventilation–perfusion matching in the lung by diverting blood flow to oxygen-rich areas. However, when alveolar hypoxia is global, as in diseases such as emphysema and cystic fibrosis, HPV leads to HPH (hypoxic pulmonary hypertension) and right heart failure. HPV is driven by the intrinsic response to hypoxia of two different cell types, namely the pulmonary arterial smooth muscle and endothelial cells. These are representatives of a group of specialized cells, commonly referred to as oxygen-sensing cells, which are defined by their acute sensitivity to relatively small changes in PO_2 and have evolved to monitor oxygen supply and alter respiratory and circulatory function, as well as the capacity of the blood to transport oxygen. Upon exposure to hypoxia, mitochondrial oxidative phosphorylation is inhibited in all such cells and this, in part, mediates cell activation. In the case of pulmonary arteries, constriction is triggered via: (i) calcium release from the smooth muscle sarcoplasmic reticulum and consequent store-depletion-activated calcium entry into the smooth muscle cells and, (ii) the modulation of transmitter release from the pulmonary artery

[1]*To whom correspondence should be addressed (email ame3@st-and.ac.uk).*

endothelium, which leads to further constriction of the smooth muscle by increasing the sensitivity of the contractile apparatus to calcium.

Introduction

Awareness of HPV (hypoxic pulmonary vasoconstriction) dawned in 1894 when Bradford and Dean [1] recorded a rise in pulmonary arterial pressure upon asphyxia. However, it was another 50 years before von Euler and Liljestrand [2] showed that hypoxia without hypercapnia induced constriction within the pulmonary circulation. In fact, von Euler and Liljestrand were the first to hypothesize that HPV may aid ventilation–perfusion matching, by diverting blood flow away from poorly ventilated areas of the lung. Thus HPV came to be recognized as the critical and distinguishing characteristic of pulmonary arteries. In contrast, systemic arteries dilate in response to tissue hypoxaemia, in order to match local perfusion to local metabolism.

HPV is not mediated by the autonomic nervous system

Investigations on the innervation of the pulmonary vasculature date back to the work of Bradford and Dean [1], which demonstrated that gross excitation of the spinal cord caused vasorelaxation within the systemic circulation, but had no effect on the pulmonary circulation. Significant progress on the basic mechanism of HPV was then provided by an investigation that suggested that HPV was a local response largely, or entirely, independent of the autonomic nervous system [3]. The limited influence on HPV of the innervation within the pulmonary vasculature was confirmed by denervation. Thus HPV was permitted after chemical sympathectomy (using 6-hydroxydopamine), surgical denervation of the carotid and aortic chemoreceptors [4] and after bilateral cervical vagotomy [5]. Most significantly, perhaps, bilateral lung transplants established that HPV remained unaffected following denervation in man [6]. We can conclude, therefore, that neither central nor local regulation of the autonomic nervous system plays a role in mediating HPV.

Airway PO_2 determines pulmonary vascular perfusion pressure at the level of pre-capillary resistance arteries

In 1951, Duke [7] demonstrated that HPV was not induced when the lung was perfused with hypoxic blood at a constant, normoxic alveolar oxygen tension. Later work [8] confirmed that the PO_2 of the perfusate was not the determining factor, but instead that a fall in alveolar PO_2 consistently triggered a pronounced increase in pulmonary vascular perfusion pressure. Clearly, therefore, HPV is triggered by a reduction in oxygen supply to the airways and/or the alveoli. This is consistent with the findings of Kato and Staub [9] who demonstrated, using unilobar hypoxia, that the small pre-capillary resistance arteries contributed most to the increase in pulmonary vascular perfusion pressure during alveolar hypoxia. Furthermore, it has been shown that the magnitude

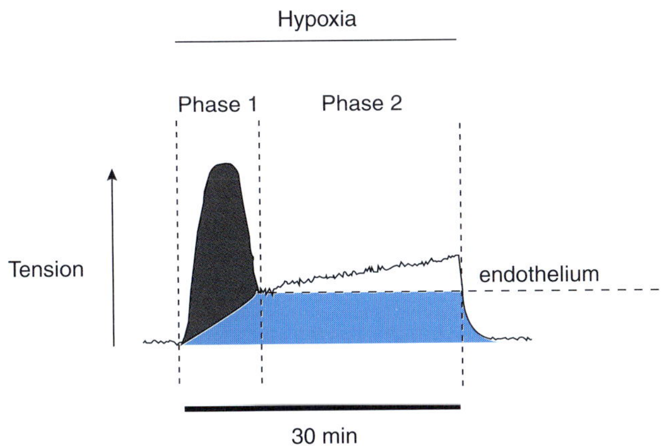

Figure 1. Hypoxic pulmonary vasoconstriction of an isolated pulmonary artery ring
Schematic diagram derived from an experimental record shows the transient Phase I constriction (black) and the sustained Phase 2 constriction in response to hypoxia. The contribution of the smooth muscle (black and blue) and endothelium (white) to HPV is also indicated.

of HPV in isolated pulmonary arteries is inversely related to pulmonary artery diameter [9]. It seems likely, therefore, that the principal mechanism(s) that drives HPV would offer a degree of selectivity for small compared with large pulmonary arteries, coupled with an even greater degree of selectivity for pulmonary versus systemic arteries.

Characteristics of the hypoxic response in isolated pulmonary arteries

In isolated pulmonary arteries, HPV is biphasic when induced by switching from a normoxic to a hypoxic gas mixture (Figure 1). Thus hypoxia induces an initial transient constriction (Phase 1) and a slow tonic constriction (Phase 2) [10,11]. Both phases of constriction are superimposed upon each other, i.e. they are initiated immediately on exposure to hypoxia. The initial transient constriction peaks within 5–10 min of the hypoxic challenge, while the underlying, tonic constriction peaks after 30–40 min [12,13]. In the isolated and perfused rat lung, however, a monophasic and sustained increase in perfusion pressure is induced by hypoxia, which has led to the physiological role of the transient, Phase 1 constriction of isolated pulmonary arteries being questioned [12,14].

HPV is mediated by mechanisms intrinsic to pulmonary arterial smooth muscle and endothelial cells

Consideration of the full facts available can leave us in no doubt that hypoxia triggers pulmonary artery constriction by activating mechanisms intrinsic to the smooth muscle and by modulating the release of mediators of constriction/dilation from the pulmonary artery endothelium (Figure 1). Thus pulmonary arterial smooth muscle and endothelial cells are now considered to be representatives of a group of specialized cells, commonly referred to as oxygen-sensing cells, which include carotid body type I cells, neuroepithelial cell bodies found in the airways of the lung and erythropoetin-secreting cells of the kidney. Such cells are defined by their acute sensitivity to relatively small changes in PO_2 and have evolved to monitor oxygen supply and alter respiratory and circulatory function, as well as the capacity of the blood to transport oxygen [15].

Inhibition of mitochondrial oxidative phosphorylation may underpin the regulation by hypoxia of oxygen-sensing cells

The precise mechanism(s) by which hypoxia elicits pulmonary artery constriction represents a more contentious issue. However, comparative studies on different oxygen-sensing cell types, including pulmonary arterial smooth muscle cells, carotid body type I cells and neuroepithelial cell bodies, have provided one consistent finding. That is that relatively mild hypoxia inhibits mitochondrial oxidative phosphorylation (for review see [16]) as indicated by depolarization of the mitochondrial membrane potential and/or by an increase in β-NADH levels, the latter having been shown to occur in isolated pulmonary arteries [17,18]. Such effects are observed over a range of PO_2 that elicits no such change in mitochondrial function in cells that do not function to monitor PO_2 and led, therefore, to the proposal that the inhibition of mitochondrial oxidative phosphorylation may underpin, at least in part, the activation of oxygen-sensing cells by hypoxia. Others have argued that the affinity of cytochrome *c* oxidase for oxygen is too high to allow for the inhibition of mitochondrial oxidative phosphorylation by physiological levels of hypoxia. However, there is clear evidence to suggest that respiratory control by oxygen may be determined by cell- and tissue-specific differences in: (i) the metabolic environment of mitochondria; (ii) the oxygen affinity of cytochrome *c* oxidase; and (iii) the dependence of the oxygen affinity of cytochrome *c* oxidase on metabolic state and rate [19].

How may the inhibition of mitochondrial metabolism by hypoxia couple to pulmonary artery constriction?

Despite extensive investigation, the mechanism(s) that may couple inhibition of mitochondrial oxidative phosphorylation to activation of oxygen-sensing cells has remained elusive. With respect to HPV there are currently three discrete hypotheses (Figure 2).

A reduction in the cellular redox status

As mentioned previously, one consequence of the inhibition of mitochondrial oxidative phosphorylation by hypoxia would be a reduction in the cellular redox status. Clearly, β-NADH levels would be expected to increase due to reduced oxidation of β-NADH at complex I of the electron transport chain

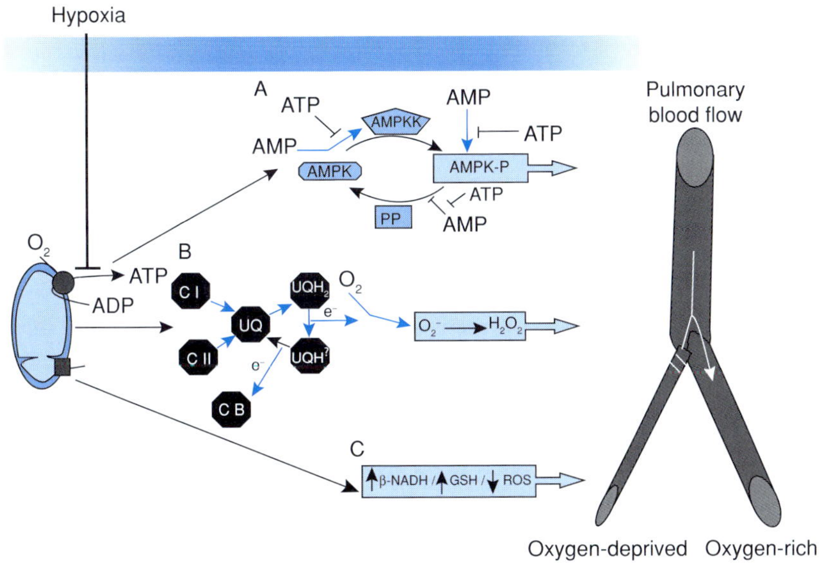

Figure 2. Mechanisms that may couple inhibition of mitochondrial oxidative phosphorylation to hypoxic pulmonary vasoconstriction

(**A**) Activation by AMPKK (AMP-activated protein kinase kinase) of AMPK in response to an increase in the cellular AMP/ATP ratio. PP, protein phosphatase. (**B**) Electron transfer and ROS production via complex III of the mitochondrial electron transport chain. C I, complex I; C II, complex II; CB, cytochrome B; UQ, ubiquinol; UQH_2, reduced ubisemiquinone; UQH˙, ubisemiquinone radical. (**C**) Reduction in the cellular redox status. GSH, reduced glutathione.

and by a consequent increase in β-NADH generation via anaerobic glycolysis. Thus hypoxia has been shown to increase β-NADH levels in pulmonary artery lysates and to increase β-NAD(P)H autofluorescence in isolated pulmonary artery rings [17,18]. Furthermore, hypoxia may increase the levels of reduced glutathione. It has been proposed, therefore, that HPV may be mediated by an increase in the levels of reduced cellular redox couples and/or by a reduction in the production of ROS (reactive oxygen species) by mitochondria, with cell signalling in response to hypoxia consequent to the reduction of sulfydryl groups on cysteine and methionine residues within key signalling proteins [20]. It should be noted, however, that although reducing agents (e.g. dithiothrietol) and some mitochondrial inhibitors (e.g. rotenone) may induce pulmonary artery constriction, investigations have yet to demonstrate that a reduced cellular redox status alone may elicit pulmonary artery constriction with the precise characteristics of HPV.

A paradoxical increase in the production of ROS

In marked contrast with the above is the proposal that a paradoxical increase in ROS may be elicited by a fall in oxygen supply. This hypothesis suggests that ROS production occurs at complex III of the mitochondrial electron transport chain, due to the uncoupled movement of electrons across the electron transport chain and their consequent interaction with molecular oxygen to form superoxide. Under normoxic conditions, complex III of the electron transport chain serves to accept electrons from ubiquinol and then transfers them to cytochrome *c* while translocating protons across the inner mitochondrial membrane. Central to this process is the Q cycle, by which complex III converts paired electron transfer from complex I and II into a sequential electron transfer that is required by complex IV. Ubiquinol, which binds to complex III at the Q_o site located at the inner mitochondrial membrane, transfers an electron to the Rieske iron-sulfur protein, for onward transfer to cytochrome *c*. This process yields the univalently reduced ubisemiquinone, which subsequently transfers a second electron to cytochrome *b* for onward transfer to a second quinone site (Q_i) at the inner mitochondrial membrane. It has been proposed that hypoxia reduces the rate of electron transfer, thereby increasing the half-life of ubisemiquinone, and that this prolonged half-life allows molecular oxygen to capture an electron from ubisemiquinone to yield superoxide even though substrate (oxygen) supply has declined. Subsequent metabolism of superoxide to H_2O_2 via mitochondrial superoxide dismutase would then increase H_2O_2 levels sufficiently to allow for diffusion through the cytoplasm and activation, via oxidation of cysteine and methionine residues in key proteins, signalling pathways that underpin HPV [21]. Evidence in support of the view that hypoxia increases ROS production and thereby induces a concomitant increase in intracellular calcium concentration has been presented [22]. However, these studies were performed on cultured pulmonary arterial smooth muscle cells, which are most likely a de-differentiated, non-contractile and proliferative phenotype and not representative of acutely isolated pulmonary artery smooth muscle cells. Furthermore, it should be noted that current techniques for the measurement of cellular ROS are notoriously difficult to perform and have provided conflicting data. Moreover, confirmatory studies in support of a role for ROS in mediating HPV in isolated pulmonary arteries have relied on the use of inhibitors of the mitochondrial electron transport chain [18] and could be interpreted as demonstrating that prior inhibition of mitochondrial oxidative phosphorylation occludes HPV without consideration of a role for ROS. However, the most significant argument against this hypothesis is the fact that acute hyperoxia, which is generally accepted to increase ROS formation, fails to induce an increase in pulmonary vascular resistance or alter the distribution of blood flow in the lung [23].

When evaluating the aforementioned hypotheses, however, it is important to consider the full extent of conflicting data now available in each case, which has precluded unification of the field behind either proposal [15].

The cellular energy status and AMPK (AMP-activated protein kinase)

Several investigations have addressed the possibility that a fall in cellular energy status may underpin HPV [24]. Assessment of the information provided by a variety of techniques has clearly shown that the cellular ATP availability/phosphorylation potential remains remarkably stable during exposure of pulmonary arteries to hypoxia, despite the fact that mitochondrial oxidative phosphorylation is inhibited. However, previous assessment of the role of the cellular energy status may have been limited by knowledge of the identity of the energy variable that might signal the response. Significantly, the role of the AMP/ATP ratio and downstream effectors had not been considered until recently, despite the fact that a variety of metabolic stresses (e.g. hypoxia, anoxia, glucose deprivation and cell activation) will precipitate an increase in the AMP/ATP ratio and that this may occur in the absence of a measurable fall in cellular ATP levels. This is due to the action of adenylate kinase, which serves to maintain cellular ATP supply during the initial phase of metabolic stress. To achieve this, adenylate kinase converts two molecules of ADP into AMP+ATP. Thus any increase in the cellular ADP/ATP ratio is converted into a much larger rise in the cellular AMP/ATP ratio. One major consequence of this is the activation of AMPK.

AMPK is a serine/threonine kinase comprising a catalytic α-subunit and regulatory β- and γ-subunits, and has come to prominence as a metabolic fuel gauge which monitors the cellular AMP/ATP ratio as an index of metabolic stress and initiates compensatory changes in cell metabolism in order to maintain ATP supply. Activation of AMPK in response to metabolic stress is most likely mediated by the binding of AMP to two Bateman (cystathionine β-synthase) domains on the γ-subunit of AMPK, which: (i) confers allosteric regulation via the γ-subunit; (ii) permits phosphorylation of the α-subunit at Thr^{172} by an upstream kinase that is a complex between the tumour suppressor kinase LKB1 and two accessory proteins STRAD (STE20-related adaptor protein) and MO25 (mouse protein 25); and (iii) inhibits dephosphorylation of AMPK. In the absence of metabolic stress, each of these processes is antagonized by high concentrations of ATP, for which the Bateman domains on the γ-subunit have lower affinity than they do for AMP. Thus AMPK is regulated by a triple mechanism that is exquisitely sensitive to very small changes in the AMP/ATP ratio.

Upon activation, AMPK promotes catabolic pathways in order to maintain ATP supply (e.g. β-oxidation of fatty acids), while switching off non-essential anabolic (ATP-consuming) pathways. Thus the primary targets for AMPK had been presumed to be genes and proteins involved in energy metabolism, particularly lipid and carbohydrate metabolism [25]. However, it is now recognised that AMPK can also target non-metabolic processes. In this respect it is important to note that while AMPK is ubiquitously expressed throughout eukaryotic cells, at least 12 different heterotrimers may be formed from multiple isoforms of the catalytic α-subunit (α1 and α2) and regulatory β- (β1 and β2) and γ- (γ1–3) subunits; splice variants of which may add to the diversity. Thus the selective expression of a particular AMPK isozyme(s) could determine, at least in part, both cell- and system specific responses to metabolic stresses. It has therefore been proposed that together with a reliance on mitochondrial oxidative phosphorylation for ATP production, the tissue-specific expression of a given AMPK heterotrimeric subunit combination could allow for the selective activation of oxygen-sensing cells in response to physiological levels of hypoxia [16].

Consistent with a role for AMPK in mediating HPV, the AMP/ATP ratio in pulmonary arterial smooth muscle has now been shown to rise 2-fold in response to hypoxia (16–21 mmHg) and vary approximately as the square of the ADP/ATP ratio, as one would expect if the adenylate kinase reaction remained close to equilibrium. Thus the AMP/ATP ratio in pulmonary artery smooth muscle may increase markedly in response to hypoxia although ATP levels may be maintained. As a consequence AMPK is activated, with catalytic activity associated with the α1 subunit isoform being selectively augmented in the acute phase. This may be significant, because under normoxic conditions the level of catalytic activity associated with the AMPKα1 subunit isoform is inversely related to pulmonary artery diameter, as is the magnitude of HPV, and is 4-fold higher in pulmonary arterial smooth muscle when compared with systemic arterial smooth muscle. Thus a specific AMPKα1-associated heterotrimeric subunit combination may provide, via signal amplification, for a degree of pulmonary selectivity that we know is required for HPV. Most significantly, pharmacological activation of AMPK has now been shown to induce pulmonary artery constriction with the precise characteristics of HPV [26], while an AMPK antagonist has been shown to inhibit acute HPV in isolated pulmonary arteries [16].

Regulation of pulmonary arterial smooth muscle by hypoxia: calcium release from the sarcoplasmic reticulum underpins smooth muscle constriction

Electrophysiological investigations have demonstrated that hypoxia inhibits the potassium current carried by K_v (voltage-sensitive potassium) channels in pulmonary arterial smooth muscle. This has led to the suggestion that HPV may be initiated by consequent membrane depolarization and voltage-gated calcium influx. However, it should be noted that although K_v channel currents were inhibited at positive membrane potentials, i.e. outside the physiological range, they were actually augmented within the range of the resting membrane potential [27]. Thus, despite the fact that other potassium channels may be inhibited by hypoxia in the appropriate membrane potential range, it is unlikely that there will be a net reduction in potassium conductance sufficient to elicit depolarization necessary for a significant increase in voltage-gated calcium influx. Despite extensive effort in this area, therefore, a contribution to acute HPV of voltage-gated calcium influx by hypoxia has yet to be established. This may explain why voltage-gated calcium channel antagonists have proved disappointing in the treatment of HPH, where their beneficial effects on pulmonary vascular resistance may be consequent to a reduction in systemic vascular resistance and cardiac output [28]. However, it should be noted that such antagonists represent the therapy of choice for hypoxic pulmonary oedema induced at high-altitude [29].

Regulation by hypoxia of calcium mobilization from sarcoplasmic reticulum calcium stores in pulmonary artery smooth muscle

Several investigations on isolated pulmonary artery smooth muscle cells and isolated pulmonary arteries have established that Phase 1 of HPV is mediated, at least in part, by the release of calcium from SR (sarcoplasmic reticulum) stores via ryanodine receptors [10]. However, there is now compelling evidence in support of a pivotal role for continued smooth muscle SR calcium release via ryanodine receptors in the maintenance of sustained HPV in isolated pulmonary arteries both with and without endothelium [10,12,30]. Briefly, both Phase 1 and Phase 2 of HPV are abolished following the block of SR calcium release via ryanodine receptors with ryanodine and caffeine [10,12], whilst constriction in response to membrane depolarization (80 mM potassium) and consequent voltage-gated calcium influx remain unaffected. Furthermore, constriction of pulmonary artery rings without the endothelium remains unaffected after removal of extracellular calcium, despite the fact that constriction induced by depolarization (80 mM potassium) is abolished. Thus it would appear that smooth muscle calcium release from a ryanodine-sensitive SR store underpins pulmonary artery smooth muscle constriction by hypoxia, rather than voltage-gated calcium entry. This view is supported by a more recent study [14] that has demonstrated that block of ryanodine receptors with ryanodine attenuates HPV in the isolated rat lung.

The calcium-mobilizing messenger cADPR (cyclic adenosine diphosphate-ribose) may mediate smooth muscle SR calcium release in response to hypoxia

cADPR is one of a family of calcium-mobilizing pyridine nucleotides and is synthesized and broken down by a multi-functional enzyme named ADP-ribosyl cyclase, which has yet to be characterized at the molecular level. An increase in cellular cADPR levels can, via an intermediate 'cADPR receptor' that remains to be identified, either increase the sensitivity of ryanodine receptors to calcium-induced calcium release or mobilize SR calcium stores via ryanodine receptor activation [31]. The fact that cADPR represents an endogenous regulator of ryanodine receptor function raised the possibility that hypoxia may mediate SR calcium release, at least in part, by increasing cADPR synthesis in pulmonary arterial smooth muscle cells.

Consistent with a role for cADPR in HPV, the vascular distribution of ADP ribosyl cyclase activity was found to offer a degree of pulmonary selectivity, being inversely related to pulmonary artery diameter, as is the magnitude of HPV [9], and at least an order of magnitude higher in homogenates of pulmonary artery smooth muscle than in those of aortic or mesenteric artery smooth muscle [30]. Furthermore, hypoxia (16–21 mm Hg) increased cADPR levels 2-fold in second order branches of the pulmonary artery, and 10-fold in third order branches [30]. Therefore the differential expression of ADP-ribosyl cyclase may offer, via signal amplification, a degree of pulmonary selectivity that is necessary for HPV. The precise mechanism(s) by which hypoxia elicits cADPR accumulation in pulmonary artery smooth muscle remains to be elucidated. However, there is evidence consistent with the view that the metabolic sensor AMPK may couple, at least in part, inhibition of mitochondrial oxidative phosphorylation to cADPR-dependent SR calcium release [26]. Furthermore, this process may be enhanced by β-NADH formation consequent to inhibition of mitochondrial oxidative phosphorylation, because β-NADH may facilitate cADPR accumulation from β-NAD^+ by augmenting ADP-ribosyl cyclase and/or inhibiting cADPR hydrolase activities [30]. It should also be noted, however, that there is evidence to suggest that an increase in cellular ROS may also augment cADPR synthesis [32]. Thus all three mechanisms that have been proposed to couple inhibition of mitochondrial function to HPV may regulate either cADPR accumulation and/or cADPR-dependent SR calcium release.

The most definitive evidence in support of a role for cADPR in mediating SR calcium release in response to hypoxia has been provided by experiments using 8-bromo-cADPR, a membrane permeant cADPR antagonist. Thus, pre-incubation of isolated pulmonary artery rings with and without endothelium, with 8-bromo-cADPR abolished the development of Phase 2 of HPV and abolished HPV in the ventilated and perfused rat lung *in situ* [12,30]. Thus cADPR-dependent SR calcium release via ryanodine receptors may be a pre-requisite for the full expression of HPV in the whole lung [12]; it is worth noting that blocking of the endothelium-dependent component of HPV by 8-bromo-cADPR may be indirect, due to the fact that an endothelium-derived vasoconstrictor released by hypoxia may mediate pulmonary artery constriction by sensitizing the contractile apparatus to calcium released from the smooth muscle SR in response to hypoxia (see below).

Surprisingly, however, 8-bromo-cADPR had no effect on the transient Phase 1 constriction of isolated pulmonary arteries, despite the fact that this component of HPV was abolished by blockage of ryanodine receptor function. Therefore it seems unlikely that the Phase 1 constriction observed in isolated pulmonary artery rings contributes to either the initiation or maintenance of the increase in pulmonary vascular perfusion pressure at the level of the whole lung, although there may be an alternate physiological consequence (see below).

Store-depletion activated calcium entry and the maintenance of HPV

Maintained smooth muscle constriction in response to hypoxia exhibits a partial dependence on transmembrane calcium influx. In this respect it is of major significance that blockage of SR calcium release via ryanodine receptors with caffeine and ryanodine, or blockage of cADPR with 8-bromo-cADPR blocks the constriction of pulmonary arteries, with or without endothelium, by hypoxia. Thus the partial dependence of smooth muscle constriction on extracellular calcium is most likely due to calcium influx via the recruitment of an SR calcium release-activated store-refilling current [33,34]. However, this process of store-release activated calcium influx is clearly consequent to the mobilization of SR calcium stores rather than being directly regulated by hypoxia [26,33,34].

Does the process underpinning Phase 1 of HPV in isolated pulmonary arteries have any physiological role?

Clearly, Phase 1 of HPV in isolated pulmonary arteries is a robust response that is dependent on calcium release from the SR via ryanodine receptors. However, the contribution of the underlying mechanisms to the hypoxia-induced increase in perfusion pressure in the whole lung is questionable, so what might its physiological function be? One possibility is that a secondary action of hypoxia may be to selectively deplete a functionally segregated SR calcium store that sits in close apposition to the plasma membrane, and which normally supports vasodilation in response to activation of adenylate cyclase-coupled receptors by releasing calcium proximal to the plasma membrane in order to initiate smooth muscle hyperpolarization via the recruitment of calcium-activated potassium channels [31]. Consistent with this proposal is the finding that β-adrenoceptor agonists have been found to be ineffective in the treatment of HPH, due to the fact that pulmonary vasodilation by this pathway is attenuated by hypoxia [31].

Regulation of the pulmonary artery endothelium by hypoxia promotes smooth muscle constriction via myofilament calcium sensitization

Removal of the pulmonary arterial endothelium or removal of extracellular calcium attenuates Phase 2 of HPV by abolishing the slow, tonic constriction that develops over ~40 min. Thus calcium influx into pulmonary arterial endothelial cells is likely to underpin the endothelium-dependent component of HPV. This appears to be mediated by a reduction in the bioavailability of endothelium-derived nitric oxide [35] and the release of an endothelium-derived vasoconstrictor. The nature of the vasoconstrictor remains controversial, although there is evidence to support a role for endothelin-1 [36]. However, more recent studies have suggested that a distinct vasoconstrictor may be the main protagonist [37]. Its nature remains to be determined, but current evidence suggests that it is heat stable and pulmonary selective. Perhaps most significantly, this vasoconstrictor does not appear to mobilize smooth muscle SR calcium stores or trigger voltage-gated calcium influx into pulmonary arterial smooth muscle. Consistent with this finding, the endothelium-dependent component of HPV develops in a manner that is uncoupled from further changes in intracellular calcium concentration in the smooth muscle. Thus it has been proposed that the principle vasoconstrictor released from the endothelium in response to hypoxia increases the sensitivity of the smooth muscle contractile apparatus to calcium [11].

In smooth muscle, the major determinant of cross-bridge cycling is the activated MLCK (myosin light chain kinase) to activated MLCP (myosin light chain phosphatase) ratio. This determines the level of phosphorylation of the MLCs (myosin light chains) and thereby the degree of contraction of the muscle. Classically, an increase in intracellular calcium concentration leads to calcium-calmodulin-dependent activation of MLCK and thereby increases the activated MLCK/activated MLCP ratio. However, it is now generally recognised that the activated MLCK/activated MLCP ratio may be increased in a calcium-independent manner by phosphorylation and consequent inhibition of MLCP. This process of MLCP phosphorylation can be induced by a number of signalling pathways that are regulated by G-protein-coupled receptors, and in this respect the monomeric G-protein RhoA and Rho-kinase (a coiled-coil-forming serine/threonine kinase) may be of primary importance in arterial smooth muscle [38].

There is now compelling evidence to support a role for the RhoA/Rho-kinase pathway in calcium sensitization during HPV. Firstly, Y-27632, a potent and highly selective Rho-kinase antagonist, inhibits both HPV in isolated rat pulmonary arteries and the hypoxic pressor response in the rat lung [39]. Most recently, hypoxia has been shown to increase Rho-kinase activity by ~260% in isolated pulmonary arteries [40]. Furthermore, in endothelium-denuded arteries the increase in Rho-kinase in response to hypoxia was much lower (~40%) and of a similar magnitude to that observed in cultured smooth muscle cells [41]. Therefore smooth muscle myofilament calcium sensitization in response to hypoxia is likely to be mediated by Rho-kinase and in an endothelium-dependent manner. It is also evident that Rho-kinase activation may represent a primary driving force for the development of hypoxic pulmonary hypertension, and that Rho-kinase antagonists may be advantageous in the treatment of this condition in humans [42]. To date, however, the complexities of this signalling pathway with respect to HPV remain to be determined. Despite this fact we can gain some insight from studies on other smooth muscle preparations (Figure 3).

RhoA is a member of a subfamily of the small monomeric GTP-binding proteins which act as molecular 'on-off' switches controlling a number of cellular processes. Like all Rho proteins, RhoA cycles between active GTP-bound and inactive GDP-bound forms. It is in the GTP-bound form that it mediates cellular processes via interactions with downstream effectors. This cycle is under the control of three groups of regulatory proteins. In the inactive GDP-bound form, RhoA is locked in the cytosol by GDIs (guanine nucleotide dissociation inhibitors), which prevent GTP binding and consequent translocation to the plasma membrane. Upon activation of G-protein-coupled receptors, GEFs (guanine-nucleotide-exchange factors), which catalyse the exchange of GDP for GTP by Rho proteins, are activated. As a consequence GTP–RhoA translocates to the plasma membrane, leading to activation of downstream effector molecules. Subsequently, inactivation of RhoA occurs via

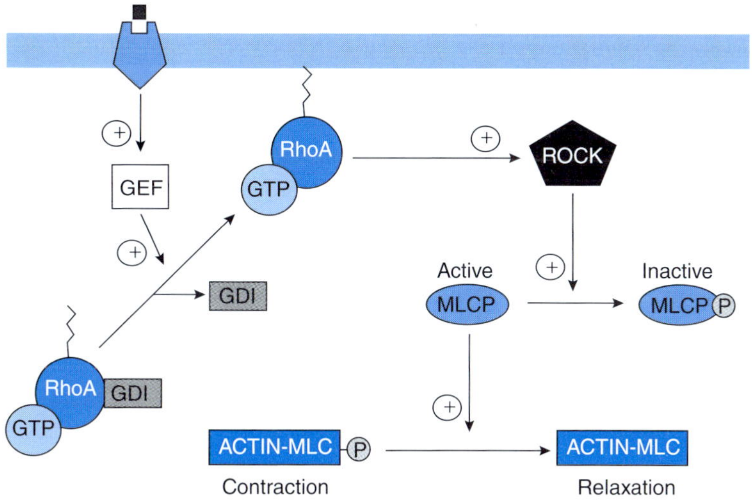

Figure 3. The RhoA/Rho-kinase pathway
Schematic representation of the processes involved in smooth muscle myofilament calcium sensitization by the monomeric G-protein RhoA and its downstream target Rho-kinase. ROCK, Rho-kinase.

its intrinsic GTPase activity, which may be accelerated via the action of GAPs (GTPase-activating proteins).

Rho-kinase is activated when GTP–RhoA binds to its C-terminus, which uncovers a catalytic phosphotransferase domain. Once activated Rho-kinase phosphorylates MLCP at its myosin-binding site. This inhibits the catalytic subunit of MLCP (PP1c) and thereby reduces dephosphorylation of MLC. Thus activation of Rho-kinase results in an increase in the activated MLCK/activated MLCP ratio and thereby sensitizes the contractile apparatus to calcium. There are, however, additional pathways via which Rho-kinase may promote myofilament calcium sensitization, for example: (i) phosphorylation of the protein kinase C substrate CPI-17, a phospho-protein which acts as an inhibitory modulator of MLCP; (ii) direct and calcium-independent phosphorylation of MLCs; and (iii) inhibition of LIM kinase, which disrupts the actin depolymerizing protein cofilin and thereby stabilizes filamentous actin [38].

Summary

- *Current evidence suggests that HPV is initiated in a manner independent of the autonomic nervous system, but dependent upon the intrinsic response of pulmonary arterial smooth muscle and endothelial cells to hypoxia.*
- *Activation of each of these cell types occurs, at least in part, as a consequence of the $P\text{O}_2$-dependent inhibition of mitochondrial oxidative phosphorylation.*

- *This may lead to the activation of AMPK combined with an increase in cellular β-NADH levels, or alternatively to a paradoxical increase in ROS production by mitochondria.*
- *Thereafter cADPR-dependent mobilization of calcium from the smooth muscle SR via ryanodine receptors is induced, together with calcium influx due to consequent activation of an SR store-refilling current. Thus pulmonary artery constriction will be initiated and maintained via calcium/calmodulin-dependent activation of MLCK.*
- *Pulmonary artery constriction may then be augmented by (i) reduced nitric oxide bioavailability; and (ii) the release of a vasoconstrictor from the pulmonary artery endothelium which may elicit smooth muscle myofilament calcium sensitization via Rho-kinase activation, leading to inhibition of MLCP and a consequent increase in the activated MLCK/activated MLCP ratio.*

References

1. Bradford, J.R. & Dean, H.P. (1894) The pulmonary circulation. *J. Physiol.* **16**, 34–158.25
2. von Euler, U.S. & Liljestrand, G. (1946) Observations on the pulmonary arterial blood pressure in the cat. *Acta Physiol. Scand.* **12**, 301–320
3. Nisell, O. (1951) The influence of blood gases on the pulmonary vessels of the cat. *Acta Physiol. Scand.* **23**, 85–90
4. Naeije, R., Lejeune, P., Leman, M., Melot, C. & Closset, J. (1989) Pulmonary vascular responses to surgical chemodenervation and chemical sympathectomy in dogs. *J. Appl. Physiol.* **66**, 42–50
5. Lejeune, P., Vachiery, J.L., Leeman, M., Brimioulle, S., Hallemans, R., Melot, C. & Naeije, R. (1989) Absence of parasympathetic control of pulmonary vascular pressure-flow plots in hyperoxic and hypoxic dogs. *Respir. Physiol.* **78**, 123–133
6. Robin, E.D., Theodore, J., Burke, C.M., Oesterle, S.N., Fowler, M.B., Jamieson, S.W., Baldwin, J.C., Morris, A.J., Hunt, S.A., Vankessel, A. et al. (1987) Hypoxic pulmonary vasoconstriction persists in the human transplanted lung. *Clin. Sci.* **72**, 283–287
7. Duke, H.N. (1951) Pulmonary vasomotor responses of isolated perfused cat lungs to anoxia and hypercapnia. *Q. J. Exper. Physiol.* **36**, 75–88
8. Bergofsky, E.H., Haas, F. & Porcelli, R. (1968) Determination of the sensitive vascular sites from which hypoxia and hypercapnia elicit rises in pulmonary arterial pressure. *Fed. Proc.* **27**, 1420–1425
9. Kato, M. & Staub, N.C. (1966) Response of small pulmonary arteries to unilobar hypoxia and hypercapnia. *Circ. Res.* **19**, 426–440
10. Dipp, M., Nye, P.C.G. & Evans, A.M. (2001) Hypoxic release of calcium from the sarcoplasmic reticulum of pulmonary artery smooth muscle. *Am. J. Physiol. Lung Cell Mol. Physiol.* **281**, L318–L325
11. Robertson, T.P., Aaronson, P.I. & Ward, J.P.T. (1995) Hypoxic vasoconstriction and intracellular Ca^{2+} in pulmonary arteries: evidence for PKC-independent Ca^{2+} sensitization. *Am. J. Physiol.* **268**, H301–H307
12. Dipp, M. & Evans, A.M. (2001) Cyclic ADP-ribose is the primary trigger for hypoxic pulmonary vasoconstriction in the rat lung *in situ*. *Circ. Res.* **89**, 77–83
13. Robertson, T.P., Hague, D., Aaronson, P.I. & Ward, J.P.T. (2000) Voltage-independent calcium entry in hypoxic pulmonary vasoconstriction of intrapulmonary arteries of the rat. *J. Physiol.* **525**, 669–680
14. Morio, Y. & McMurtry, I.F. (2002) Ca(2+) release from ryanodine-sensitive store contributes to mechanism of hypoxic vasoconstriction in rat lungs. *J. Appl. Physiol.* **92**, 527–534

15. Gonzalez, C., Sanz-Alfayate, G., Agapito, M.T., Gomez-Nino, A., Rocher, A. & Obeso, A. (2002) Significance of ROS in oxygen sensing in cell systems with sensitivity to physiological hypoxia. *Respir. Physiol. Neurobiol.* **132**, 17–41
16. Evans, A.M. (2006) AMP-activated protein kinase and the regulation of Ca^{2+} signalling in O_2-sensing cells. *J. Physiol.* **574**, 113–123
17. Shigemori, K., Ishizaki, T., Matsukawa, S., Sakai, A., Nakai, T. & Miyabo, S. (1996) Adenine nucleotides via activation of ATP-sensitive K^+ channels modulate hypoxic response in rat pulmonary artery. *Am. J. Physiol.* **270**, L803–L809
18. Leach, R.M., Hill, H.M., Snetkov, V.A., Robertson, T.P. & Ward, J.P. (2001) Divergent roles of glycolysis and the mitochondrial electron transport chain in hypoxic pulmonary vasoconstriction of the rat: identity of the hypoxic sensor. *J. Physiol.* **536**, 211–224
19. Gnaiger, E., Lassnig, B., Kuznetsov, A., Rieger, G. & Margreiter, R. (1998) Mitochondrial oxygen affinity, respiratory flux control and excess capacity of cytochrome c oxidase. *J. Exp. Biol.* **201**, 1129–1139
20. Michelakis, E.D., Thebaud, B., Weir, E.K. & Archer, S.L. (2004) Hypoxic pulmonary vasoconstriction: redox regulation of O_2-sensitive K^+ channels by a mitochondrial O_2-sensor in resistance artery smooth muscle cells. *J. Mol. Cell. Cardiol.* **37**, 1119–1136
21. Guzy, R.D. & Schumacker, P.T. (2006) Oxygen sensing by mitochondria at complex III: the paradox of increased reactive oxygen species during hypoxia. *Exp. Physiol.* **91**, 807–819
22. Waypa, G.B., Guzy, R., Mungai, P.T., Mack, M.M., Marks, J.D., Roe, M.W. & Schumacker, P.T. (2006) Increases in mitochondrial reactive oxygen species trigger hypoxia-induced calcium responses in pulmonary artery smooth muscle cells. *Circ. Res.* **99**, 970–978
23. Hambraeus-Jonzon, K., Bindslev, L., Mellgard, A.J. & Hedenstierna, G. (1997) Hypoxic pulmonary vasoconstriction in human lungs. A stimulus-response study. *Anesthesiology* **86**, 308–315
24. Leach, R.M., Sheehan, D.W., Chacko, V.P. & Sylvester, J.T. (2000) Energy state, pH, and vasomotor tone during hypoxia in precontracted pulmonary and femoral arteries. *Am. J. Physiol.* **278**, L294–L304
25. Hardie, D.G. (2004) AMP-activated protein kinase: a master switch in glucose and lipid metabolism. *Rev. Endocr. Metab. Disord.* **5**, 119–125
26. Evans, A.M., Mustard, K.J., Wyatt, C.N., Peers, C., Dipp, M., Kumar, P., Kinnear, N.P. & Hardie, D.G. (2005) Does AMP-activated protein kinase couple inhibition of mitochondrial oxidative phosphorylation by hypoxia to calcium signaling in O_2-sensing cells? *J. Biol. Chem.* **280**, 41504–41511
27. Post, J., Hume, J., Archer, S. & Weir, E. (1992) Direct role for potassium channel inhibition in hypoxic pulmonary vasoconstriction. *Am. J. Physiol.* **262**, C882–C890
28. Neely, C.F., Stein, R., Matot, I., Batra, V. and Cheung, A. (1996) Calcium blockage in pulmonary hypertension and hypoxic vasoconstriction. *New Horiz.* **4**, 99–106
29. Bartsch, P., Maggiorini, M., Ritter, M., Noti, C., Vock, P. & Oelz, O. (1991) Prevention of high-altitude pulmonary edema by nifedipine. *N. Engl. J. Med.* **325**, 1284–1289
30. Wilson, H.L., Dipp, M., Thomas, J.M., Lad, C., Galione, A. & Evans, A.M. (2001) ADP-ribosyl cyclase and cyclic ADP-ribose hydrolase act as a redox sensor: a primary role for cADPR in hypoxic pulmonary vasoconstriction. *J. Biol. Chem.* **276**, 11180–11188
31. Evans, A.M., Wyatt, C.N., Kinnear, N.P., Clark, J.H. & Blanco, E.A. (2005) Pyridine nucleotides and calcium signalling in arterial smooth muscle: from cell physiology to pharmacology. *Pharmacol. Ther.* **107**, 286–313
32. Zhang, A.Y., Yi, F., Teggatz, E.G., Zou, A.-P. & Li, P.-L. (2004) Enhanced production and action of cyclic ADP-ribose during oxidative stress in small bovine coronary arterial smooth muscle. *Microvasc. Res.* **67**, 159–167
33. Kang, T.M., Park, M.K. & Uhm, D.Y. (2003) Effects of hypoxia and mitochondrial inhibition on the capacitative calcium entry in rabbit pulmonary arterial smooth muscle cells. *Life Sci.* **72**, 1467–1479
34. Ng, L.C., Wilson, S.M. & Hume, J.R. (2005) Mobilization of sarcoplasmic reticulum stores by hypoxia leads to consequent activation of capacitative Ca^{2+} entry in isolated canine pulmonary arterial smooth muscle cells. *J. Physiol.* **563**, 409–419

35. Weissmann, N., Winterhalder, S., Nollen, M., Voswinckel, R., Quanz, K., Ghofrani, H.A., Schermuly, R.T., Seeger, W. & Grimminger, F. (2001) NO and reactive oxygen species are involved in biphasic hypoxic vasoconstriction of isolated rabbit lungs. *Am. J. Physiol. Lung Cell Mol. Physiol.* **280**, L638–L645
36. Liu, Q., Sham, J.S., Shimoda, L.A. & Sylvester, J.T. (2001) Hypoxic constriction of porcine distal pulmonary arteries: endothelium and endothelin dependence. *Am. J. Physiol. Lung Cell Mol. Physiol.* **280**, L856–L865
37. Robertson, T.P., Ward, J.P. & Aaronson, P.I. (2001) Hypoxia induces the release of a pulmonary-selective, Ca^{2+}-sensitising, vasoconstrictor from the perfused rat lung. *Cardiovasc. Res.* **50**, 145–150
38. Somlyo, A.P. & Somlyo, A.V. (2003) Ca^{2+} sensitivity of smooth muscle and nonmuscle myosin II: modulated by G proteins, kinases, and myosin phosphatase. *Physiol. Rev.* **83**, 1325–1358
39. Robertson, T.P., Dipp, M., Ward, J.P.T., Aaronson, P.I. & Evans, A.M. (2000) Inhibition of sustained hypoxic vasoconstriction by Y-27632 in isolated intrapulmonary arteries and perfused lung of the rat. *Br. J. Pharmacol.* **131**, 5–9
40. Robertson, T., Hashmi-Hill, M., Vandenplas, M.L. & Lewis, S.J. (2005) Endothelium-dependent activation of Rho-kinase during hypoxic pulmonary vasoconstriction in rat intrapulmonary arteries. *FASEB J.* **19**, A1277
41. Wang, Z., Jin, N., Ganguli, S., Swartz, D.R., Li, L. & Rhoades, R.A. (2001) Rho-kinase activation is involved in hypoxia-induced pulmonary vasoconstriction. *Am. J. Respir. Cell Mol. Biol.* **25**, 628–635
42. Ishikura, K., Yamada, N., Ito, M., Ota, S., Nakamura, M., Isaka, N. & Nakano, T. (2006) Beneficial acute effects of rho-kinase inhibitor in patients with pulmonary arterial hypertension. *Circ. J.* **70**, 174–178

6

Oxygen sensing by ion channels

Paul J. Kemp*[1] and Chris Peers†

**Cardiff School of Bioscience, Cardiff University, Museum Avenue, Cardiff CF10 3US, U.K. and †School of Medicine, Worsley Building, University of Leeds, Leeds LS2 9JT, U.K.*

Abstract

The ability to sense and react to changes in environmental oxygen levels is crucial to the survival of all aerobic life forms. In mammals, specialized tissues have evolved which can sense and rapidly respond to an acute reduction in oxygen and central to this ability in many is dynamic modulation of ion channels by hypoxia. The most widely studied oxygen-sensitive ion channels are potassium channels but oxygen sensing by members of both the calcium and sodium channel families has also been demonstrated. This chapter will focus on mechanisms of physiological oxygen sensing by ion channels, with particular emphasis on potassium channel function, and will highlight some of the consensuses and controversies within the field. Where data are available, this chapter will also make use of information gleaned from heterologous expression of recombinant proteins in an attempt to consolidate what we know currently about the molecular mechanisms of acute oxygen sensing by ion channels.

Introduction

A characteristic common to all cells is the ability to drive and maintain ionic gradients across their plasma membranes. The primary point of gradient generation is the ubiquitously expressed Na^+/K^+-ATPase (the 'sodium pump'),

[1]*To whom correspondence should be addressed (email kemp@cf.ac.uk).*

which is responsible for pumping out of the cell three sodium ions in exchange for two potassium ions during a single enzymatic cycle as one molecule of ATP is hydrolysed. The continued and tightly regulated activity of this pumping system generates opposing sodium and potassium gradients across the plasma membrane of cells, the potential energy of which is harnessed by numerous integral plasma membrane transporters and ion channels in order to co-ordinate the cell-specific processes which define cellular phenotype. Of particular note, and common to all eukaryotic cells, is the rapid diffusion of potassium across the plasma membrane through potassium-selective channels, with the physiochemical consequence that a negative transmembrane potential develops at rest, termed the E_m (resting membrane potential). The magnitude of the resting potassium permeability (in the absence of any marked change in the potassium gradient) is the single most important regulator of E_m in most excitable and non-excitable cells. It follows therefore, that any factors which directly affect cellular potassium permeability will have a dramatic influence on the E_m and excitability of cells. One factor which has emerged in recent years as an important regulator of potassium permeability is the availability of molecular oxygen.

Excitable cells, by definition, do more than maintain their E_m at between −40 and −80 mV. Rather, they have the ability to generate action potentials, a process which is characterized by a rapid loss of their negative E_m as the action potential upstroke is generated by the rapid diffusion of sodium into the cell down its chemical gradient through specialized, voltage-dependent sodium channels. This drives the E_m briefly to potentials approaching the equilibrium potential for sodium. Such cell depolarization increases the membrane permeability of a third cation, calcium. Influx through VGCCs (voltage-gated calcium channels) has a profound effect on numerous cellular processes including vesicular neurotransmitter release and muscle contraction. Cells then repolarize as voltage-dependent sodium channels close through inactivation and voltage-dependent potassium channels open, promoting potassium flow out of the cell again. As for the E_m, it follows that factors affecting this orchestrated opening and closure of voltage-dependent ion channels will have a dramatic impact on the electrical characteristics of excitable cells and their biochemical sequelae. Again, one such factor is the availability of oxygen.

To date, the activity of specific members of almost all ion channel families have been shown experimentally to be regulated by acute changes in the PO_2 (partial pressure of oxygen). However, in many cases (such as central neuronal potassium channels, neuronal sodium channels and cardiac sodium channels) these effects are likely to reflect a pathological response to oxygen deprivation, such as is experienced in stroke or myocardial infarction, and so will not be discussed in detail here (but see [1] for further details). Instead, this chapter focuses primarily on proposed mechanisms of physiological oxygen-sensing by ion channels and outlines the consensuses and controversies which surround a number of worked examples. In so doing, this draws on data obtained from ion channels expressed natively in tissues classically thought of as oxygen-sensing, whose function it is to regulate whole organism oxygen levels (such as the carotid body and vascular smooth muscle). In addition, data are drawn from tissues not normally associated directly with oxygen-sensing, but whose responses to hypoxia can still be considered as a physiological rather than a pathological phenomenon (such as T-lymphocytes). Where available, it also makes use of information gleaned from heterologous expression of recombinant proteins in an attempt to consolidate what we currently know about the molecular mechanisms underlying the important process of acute oxygen sensing by ion channels.

The role of oxygen-sensing ion channels in physiology

The first definitive demonstration of acute oxygen-sensing by any ion channel came from the ground-breaking work of Lopez-Barneo et al. [2]. This seminal work demonstrated the reversible inhibition of a specific potassium conductance in rabbit carotid body glomus cells. Such potassium channel inhibition was proposed to account for glomus cell depolarization, voltage-dependent calcium influx and consequent neurotransmitter release during hypoxia. Thus hypoxic inhibition of potassium channels is central to the carotid body chemoreflex that initiates the increased ventilatory response in the face of reduced arterial PO_2. In addition to the carotid body-derived response to systemic hypoxaemia, airway hypoxia evokes vasoconstriction in the pulmonary vascular bed, thus shunting blood away from poorly ventilated alveoli. Although controversial, there is a large body of evidence to suggest that oxygen-sensing, particularly by potassium channels, is central to this vasoconstrictor response (see Chapter 5). Even more contentious is a role for pulmonary neuroepithelial bodies of the lung in the oxygen-sensing. Neuroepithelial bodies (and cellular models thereof) certainly express oxygen-sensing potassium channels and release serotonin during hypoxia. Furthermore, they exhibit an innervation pattern characteristic of other pulmonary receptors even though their physiological function remains somewhat elusive. Oxygen-sensing potassium channels have also been implicated in processes as diverse as control of catecholamine release from the fetal adrenal medulla and T-lymphocyte activation. As hypoxia is evoking ventilatory compensation and pulmonary vasoconstriction, several other vascular beds begin to dilate in order to optimize oxygen delivery to important vascular beds such as the cerebral and coronary circulations; such vasodilatation is dependent upon hypoxic inhibition of a specific member of the VGCC family.

Oxygen-sensing by potassium channels

Potassium channels represent the largest and most diverse of the ion channel families, but can be subdivided according to the structure of their pore-forming α-subunits. The most commonly described are represented in Figure 1, alongside a list of oxygen-sensitive examples which are dealt with in more detail below.

Predicted topology

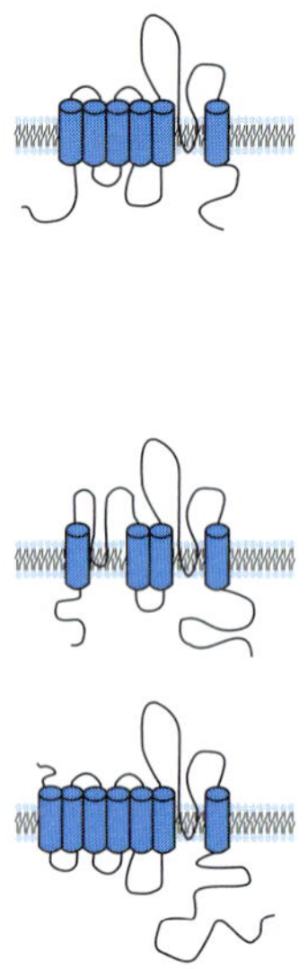

Nomenclature

HGNC	IUPHAR	Other
KCNA1	Kv1.1	Shaker
KCNA2	Kv1.2	Shaker
KCNA3	Kv1.3	Shaker
KCNA5	Kv1.5	Shaker
KCNB1	Kv2.1	Shab
KCNC1	Kv3.1b	Shaw
KCNC3	Kv3.3b	Shaw
KCND1	Kv4.1	Shal
KCND3	Kv4.3	Shal
KCNK3	TASK-1	K_{2P}
KCNK2	TREK-1	K_{2P}
KCNK9	TASK-3	K_{2P}
KCNK13	THIK-1	K_{2P}
KCNMA1	$K_{Ca}1.1$	BK_{Ca}, maxi-K, slo

Figure 1. Known oxygen-sensitive potassium channels
The cartoon depictions at the left of the figure represent generic and predicted topologies of the three known families of potassium channel proteins known to be inhibited by acute hypoxia. On the right of the figure the molecular identity of each individual potassium channel is shown along with the three methods of classification. These are, from left to right, the HGNC (Human Genome Name Classification), the IUPHAR (International Union of Pharmacology) classification and the initial drosophila classification (Other).

Shaker channel family

Oxygen-sensing by members of the *shaker* family of potassium channels (termed KCNAx) has been implicated in the physiological responses to hypoxia in the carotid body, pulmonary arterioles, T-lymphocytes and chromaffin cells. KCNA1 (Kv1.1) null mice show a greater hypoxic ventilatory response and larger carotid sinus nerve discharge during hypoxia than do the wild-type controls. KCNA1 is expressed in the carotid body, petrosal ganglion cells and the nucleus of the tractus solitarius, all constituent parts of the afferent limb of peripheral chemosensory reflexes. The molecular nature of this hypersensitivity is not clear but since there has been no demonstration of 'innate' oxygen sensitivity of KCNA1, the simplest explanation is that its absence may simply evoke an increase in excitability by bringing chemosensory cells closer to threshold [3]. KCNA2 (Kv1.2) has been shown to be oxygen-sensitive in PC12 cells [4] and, when expressed in combination with KCNA5 (Kv1.5), is inhibited by hypoxia in pulmonary arteriolar smooth muscle [5]. KCNA3 (Kv1.3) is involved in T-lymphocyte activation. When co-expressed with the Src tyrosine kinase, Lck, KCNA3 becomes oxygen-sensitive and is inhibited by hypoxia, in part explaining why lymphocyte proliferation/activation is impaired in low oxygen environments, such as tumours [6]. KCNA5 is oxygen-sensitive when overexpressed in pulmonary arterial myocytes [although not in mesenteric arterial myocytes or HEK (human embryonic kidney) cells] suggesting that a pulmonary-specific interaction with this channel is required to confer oxygen sensitivity [7]. KCNA6–10 have not been shown to be sensitive to acute hypoxia although there is a suggestion that a splice variant of KCNA7 may have the potential for such sensing in the heart [8].

Shab channel family

Of the two known *shab* potassium channel members (termed KCNBx), only KCNB1 (Kv2.1) has been shown to be oxygen sensitive in certain tissues and cell types. Although there are reports that this channel is inhibited by hypoxia in pulmonary smooth muscle [9], it seems likely that it requires co-expression of the silent KCNS3 (Kv9.3) subunit [5,10].

Shaw channel family

KCNC1 (Kv3.1b) has been suggested as the main target of hypoxic inhibition in pulmonary arteriolar smooth muscle by a number of investigators [11,12]. Recombinant KCNC3 (Kv3.3b) demonstrates oxygen sensitivity and this is maintained when natively expressed in the putative airway oxygen-sensing tissue, the neuroepithelial bodies [13], where the upstream sensing system requires free radical generation (in this case hydrogen peroxide) by the membrane-localized NADPH oxidase [14]. KCNCx family members are the primary oxygen-sensing potassium channels in the mouse carotid body [15], showing that species (or even strain) differences are important when considering the integrated effect of hypoxia on a particular tissue or reflex (see below for information regarding oxygen-sensitive potassium channel identities in the carotid bodies of other species).

Shal channel family

To date, oxygen-sensing by *shal* potassium channel members (termed KCNDx) has been only demonstrated in the rabbit carotid body. Inhibition by hypoxia of KCND1 (Kv4.1) and KCND3 (Kv4.3), either as homomers or heteromers, appears to account for the rabbit glomus cell chemosensitivity [16]. Although KCNC channel members have been excluded from the list of oxygen-sensitive channels in the carotid body isolated from normal rabbits, a recent report [17] suggests that congestive heart failure evokes a remodelling of the glomus cell potassium channel expression pattern. The resultant differential down-regulation of KCNC4 (Kv3.4) results in chronic glomus cell depolarization, which amplifies the effect of hypoxic inhibition of KNCD1/3 and explains the consequent augmentation of the hypoxic chemoreflex in congestive heart failure [17].

Kv β-subunits

It has long been appreciated that activity of the pore-containing Kv α-subunits (KCNA-D, above) is known to be modulated by co-expression with Kv β-subunits (Kvβx). Such modulatory activity may well include the conference of oxygen-sensitivity to Kv channels. For example, recombinant KCND2 (Kv4.2) is only able to respond to hypoxia when it is co-expressed with Kvβ1.2 [18] whilst Kvβ1.2 and Kvβ1.3 confer oxygen-sensitivity to KCNC1 in the pulmonary vasculature [11].

Tandem P domain potassium channel family

Members of the tandem P domain potassium channel family (termed KCNKx) are major determinants of E_m and regulators of cellular excitability as they are characteristically open at potentials around the E_m. Recombinant KNCK3 (also known as TASK1) is inhibited by hypoxia in HEK-293 cells [19] and a native KCNK3-like current has been implicated in oxygen-sensing of the carotid body [20]. KCNK3, perhaps in combination with KCNK9 (also known as TASK3), underlies the oxygen-sensitive current in the immortalized cellular counterpart of the neuroepithelial body, H146 cells [21]. That KCNK3 almost certainly requires a subsidiary oxygen-sensing system to respond to hypoxia is evidenced by the cell-specificity of its hypoxic inhibition [22], a notion strongly supported in H146 cells [23], native neuroepthelial bodies [14] and recombinant systems [17], where it has been shown that the oxygen sensitivity of this channel is conferred by the activity of an NADPH oxidase enzyme which contains either NOX2 or NOX4 subunits. KCNK3 has also been suggested to play a role in hypoxic pulmonary vascular constriction [24] and hypoxic excitotoxicity of central neurones [25], although the precise mechanisms of action are currently unknown. A further candidate for control of central neuronal excitability is KCNK2 (also known as TREK-1) which is oxygen-insensitive in some circumstances [26] and oxygen-sensitive in others [27]. Where hypoxic inhibition is apparent, it appears dependent upon structural elements within the C-terminus [28]. Recombinant KCNK13 (also known as THIK-1) is inhibited by hypoxia [29] and is expressed in efferent fibres innervating the carotid body [30], suggesting a role in modulation of the chemoreflex.

Large conductance, calcium-activated potassium channels

These channels are variously known as KCNMAx, slo, BK_{Ca} and Maxi-K channels. The first demonstration of oxygen-sensing by this channel family came from patch-clamp studies in the rat carotid body glomus cells where hypoxia inhibits the channels [31]. Subsequently, it was demonstrated that both native (carotid body-derived [32]) and recombinant (KCNMA1 [33]) forms of this channel were inhibited by hypoxia in isolated membrane patches but perhaps require another protein to confer full hypoxic inhibition.

Other potassium channels

Several members of the KCNNx (termed SK) channel family have been implicated in hypoxic signal transduction. Hypoxic inhibition of KCN2 (SK2) in prenatal adrenal medulla may contribute to fetal catecholamine release during labour [34]; the postnatal tissue is oxygen-insensitive, presumably due to tonic inhibitory influences of adrenal innervation after birth. Oxygen-sensing by a members of the KCNJx (inward rectifiers or Kir) appears to be restricted to the coronary arterial smooth muscle where hypoxia results in current activation via increased production of cAMP and vessel dilatation [35]. This is the only potassium channel known to be activated by hypoxia and such activation has only been seen in a non-pulmonary setting, again suggesting that sensitivity of potassium channels to hypoxia is a tissue-specific phenomenon and serves to reinforce the idea that the nature of the cell-specific protein partners (close or distant from the channel itself) are key to understanding responses to reduced PO_2.

Mechanisms of hypoxic channel inhibition

Although a picture has slowly been emerging concerning the increasingly widespread physiological roles of oxygen-sensitive potassium channels in various tissues, the identification of mechanisms to account for such observations has proved more elusive, and certainly more contentious (see [36] for a recent review). Several mechanisms have been proposed, as shown schematically in Figure 2, originating with the idea that redox modulation was key to potas-

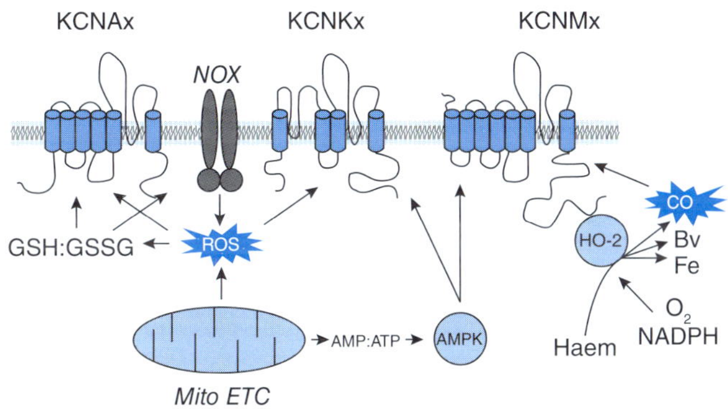

Figure 2. Principal proposed mechanisms of oxygen-sensing by potassium channels
Each channel family is depicted by its proposed topology. Haemoxygenase-2 (HO-2) produces biliverdin (BV), iron (Fe) and carbon monoxide (CO) in the presence of molecular oxygen and NADPH. Carbon monoxide is a tonic activator of BK_{Ca} (KCNMx) and its reduction during hypoxia results in channel closure. The multi-component enzyme complex NADPH oxidase (NOX) produces ROS in normoxia. During hypoxia, ROS production is reduced which evokes either direct channel closure (of KCNKx and KCNA/B/C/D) or produces an indirect effect via augmented GSH:GSSG ratio. Alternatively, ROS from mitochondria may rise and cause channel closure. Reduced mitochondrial electron transport (mito ETC) activity augments the AMP:ATP ratio, activates AMPK and closes channels via AMPK-dependent phosphorylation.

sium channel inhibition by hypoxia. Put simply, this idea suggests that levels of ROS (reactive oxygen species) vary according to oxygen levels, and thereby alter channel activity either directly, or via changes in redox-sensitive molecules such as glutathione. However, there are fundamental issues concerning this idea that currently remain unresolved.

First, what happens to ROS when cells are exposed to hypoxia? This seems a straightforward question, but is highly contested, particularly in the study of hypoxic pulmonary vasoconstriction, where there are compelling arguments that ROS decrease in hypoxia [37], yet substantial alternative data suggest ROS increase in hypoxia [38]. Secondly, what is, or are, the intracellular sources of ROS? Even those authors who disagree about the effects of hypoxia on ROS come together in the belief that oxygen-dependent ROS production originates from mitochondria. However, other sources have been proposed. In particular, hypoxic reduction of ROS by NADPH oxidase has been proposed as a mechanism by which hypoxia inhibits potassium channels in airway neuroepithelial bodies [13,23], a finding which is strongly supported by lack of potassium current inhibition in NADPH oxidase-deficient transgenic mice [14]. However, carotid body chemoreception and hypoxic pulmonary vasoconstriction continues unabated in these mice. To add to the confusion, there is good evidence that hypoxic pulmonary vasoconstriction does not rely on potassium channel inhibition (or, at least, on the voltage-gated calcium entry arising from channel inhibition), but instead depends on release of calcium from intracellular stores. This concept (see Chapter 5) again puts mitochondria in the spotlight as the key to oxygen-sensing, but not via altered ROS production. Instead, subtle changes in the cytoplasmic AMP:ATP ratio have been proposed to activate AMPK (AMP-activated protein kinase), thereby initiating hypoxic pulmonary vasoconstriction. Indeed, AMPK activation may also be important in carotid body chemotransduction since its pharmacological activation mimicked the effects of hypoxia to cause voltage-gated calcium entry [39].

Of the potassium channels expressed in rat carotid body glomus cells, both BK_{Ca} and TASK-like channels demonstrate oxygen sensitivity. The human homologue of the BK_{Ca} channel (termed KCNMA1) is closely coupled to haemoxygenase-2. This enzyme breaks down cellular haem, in the presence of NADPH and molecular oxygen, to produce carbon monoxide, biliverdin and iron. Carbon monoxide is a remarkably potent activator of recombinant [40] and native [32] BK_{Ca} channels of the carotid body and systemic arteriolar myocytes [41]. It appears that the channel-associated haemoxygenase-2 (in both recombinant systems and native glomus cells) confers an increased oxygen sensitivity to the BK_{Ca} channel via acute regulation of the channel activity by its downstream products. Thus, as oxygen availability decreases, carbon monoxide and biliverdin production wanes and channels close [40]. These data concur with those obtained *in vitro* from sinus nerve recordings in the rat (where nerve activity is augmented by blockers of haemoxygenase [42]) and *in vivo* from respiratory studies in haemoxygenase-2 knockout mice (where the hypoxic ventilatory response is blunted [43]). However, recent data obtained in carotid body slices from haemoxygenase-2 knockout mice have suggested that neurotransmitter release during hypoxia is unaffected by haemoxygenase-2 gene deletion [44]. It is not clear why these studies are different from each other although it may be pertinent that the mouse carotid body primarily uses KCNCx channels (not KCNMx channels) to sense hypoxia [15], and even this species-specificity is not consistent amongst different mouse strains [45].

At present, it seems reasonable to conclude that multiple mechanisms may exist to account for modulation of potassium (and other) ion channels by hypoxia (Figure 2). Indeed, more than one mechanism may be present in the same cell. However, this field clearly requires further exploration, and some fundamental issues remain to be resolved. In addition, clarification is needed concerning the role of specific mechanisms in physiologically relevant oxygen-sensing processes. There is clearly a long way to go.

Structural requirements for oxygen-sensing by ion channels

Despite the widespread interest in mechanisms of oxygen-sensing by ion channels, little attention has been paid to the structural requirements of ion channels for such sensing. This is perhaps surprising, given the wealth of structure/function studies in the literature. Nevertheless, some interesting advances have been made. Whilst the research community was examining the role of hypoxic potassium channel inhibition in mediating pulmonary vasoconstriction, the intrinsic oxygen-sensing properties of systemic vascular smooth muscle (which dilates under hypoxia) were being investigated. Lopez-Barneo and co-workers [46] reported that L-type calcium channels were reversibly inhibited in isolated smooth muscle cells from various systemic vascular beds in a manner that was voltage-dependent and associated with a slowing of activation kinetics. This work was rapidly extended to account not only for systemic vasodilation, but also the dilating effects of hypoxia on pulmonary conduit (rather than resistance) vessels [47,48]. Shortly following these reports, an indistinguishable effect was reported on recombinant human L-type calcium channels expressed in HEK-293 cells [49]. Importantly, this study [49] employed the major pore-forming, voltage-sensing α_{1C} subunit of the calcium channel (Cav1.2, CACNA1C), in the absence of auxiliary subunits. This immediately discounted an important role for other subunits of the calcium channel complex. Subsequent exploration of the mechanisms underlying hypoxic inhibition of calcium channels have not been forthcoming. However, insight into the structural requirements for oxygen-sensing have arisen from comparison of the three splice variants of this channel: variation occurs at a single site in the cytoplasmic C-terminal domain of the channel (Figure 3A), where an insert of 71 residues is either omitted or replaced with a shorter, structurally unrelated insert. Of these, only channels possessing the 71 amino acid insert were sensitive to hypoxia (Figure 3A).

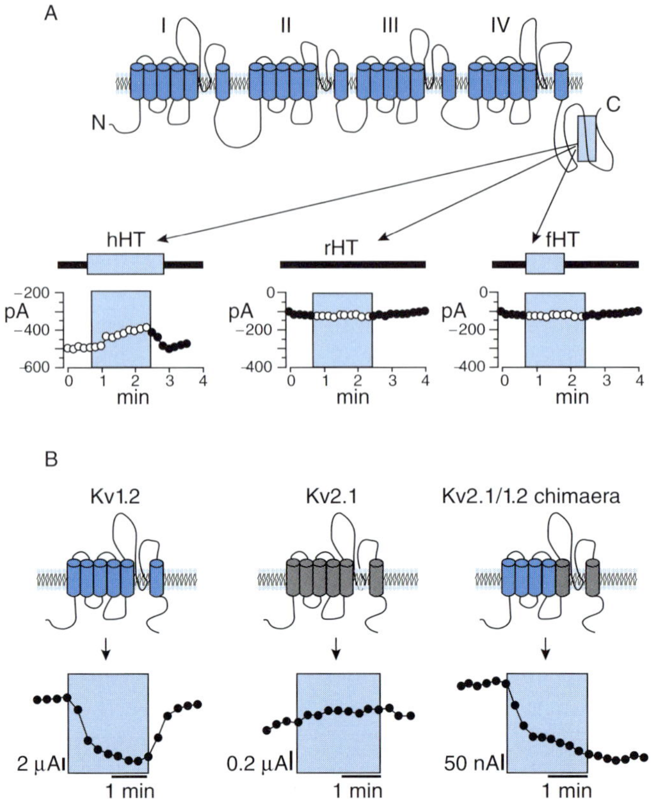

Figure 3. Structural requirements for oxygen-sensing by two ion channels
(**A**) Cartoon of the four domains (I–IV) of the α-subunit of a voltage-gated calcium channel, each consisting of six transmembrane α-helicies. The boxed area (shaded blue) in the cytoplasmic C-terminal domain represents the region of splice variation. Below the schematic of each variant is a time series plot, illustrating the calcium current amplitude evoked by successive step depolarizations from −80mV to +10mV (100 ms duration, 0.1 Hz). For the period indicated by shading, the cell (in each case, a HEK-293 cell transfected with the relevant splice variant) was exposed to a reduction of PO_2 from ~150 mmHg to ~20 mmHg. Reproduced from [51] Fearon, I.M., Varadi, G., Koch, S., Isaacsohn, I., Ball, S.G. and Peers, C. (2000) Splice variants reveal the region involved in oxygen sensing by recombinant human L-type Ca^{2+} channels. Circulation Research **87**, 537–539, with permission from Lippincott, Williams and Wilkins. (**B**) Cartoon of the predicted topology of two potassium channels, KCNA2 (Kv1.2; left, blue) and KNCB1 (Kv2.1; centre, grey) along with a chimaera of the two channels (right; relevant sections shaded accordingly). Below each cartoon is a time series plot, illustrating the potassium current amplitude evoked by successive step depolarizations from −80 mV to +50 mV (applied every 15 s). For the periods indicated by shading, the cell (in each case, a *Xenopus* oocyte, injected with the appropriate cRNA 24 h previously) was exposed to anoxia (0 mmHg oxygen). Reprinted from [51] Biochem. Biophys. Res. Commun. **306**, Conforti, L., Takimoto, K., Petrovic, M., Pongs, O. and Millhorn, D., The pore region of the Kv1.2α subunit is an important component of recombinant Kv1.2 channel oxygen sensitivity, 450–456, (2003), with permission for Elsevier.

Mutagenesis studies revealed that the oxygen-sensing region resided in the proximal 39 residues of this insert [50].

At present, such information has not been applied systematically to studies of structural requirements of potassium channels for oxygen-sensing. However, one study has employed chimaeras to reveal that oxygen-sensitivity of recombinant, homomeric KCNA2 is dependent upon an interaction within the channel pore region (see Figure 3B). Thus, oxygen-sensitivity was conferred to the apparently oxygen-insensitive homomeric KCNB1 (Kv2.1) by splicing into it the S5–S6 region of KCNA2 [51]. This study also revealed that a redox-sensitive methionine residue within the oxygen-sensing region of Kv1.2 was not involved in hypoxic responses. However, it should be noted that hypoxic inhibition of the Kv1.2/2.1 chimaera was, unlike inhibition of Kv1.2 itself, irreversible. Nevertheless, this study remains the most detailed investigation of structural requirements of potassium channels for oxygen-sensing to date.

Conclusion

The ability of potassium and calcium channels to sense and react to acute alteration in PO_2 is central to the homoeostatic function of cell-types as diverse as T-lymphocytes, vascular myocytes and neurones. The expression of oxygen-sensitive ion channels and the molecular mechanism(s) underpinning how each responds to acute hypoxia is both species and cell-specific. Moreover, it is becoming increasingly clear that the ability to sense oxygen is bestowed upon many ion channels by cell-specific protein partnerships and/or functional association with cell signalling cascades. Clearly, discovering how differential expression of these microenvironments and signalling networks is controlled within each cell type represents a significant challenge in the 21st century. That said, the ability of a particular cell-type to respond appropriately to an acute hypoxic stimulus is crucial to survival and it is, therefore, no surprise that a significant redundancy might exist in the oxygen-sensing systems. In that way, organisms will always have at least one mechanism that can be relied upon to ensure optimal performance during periods of oxygen deprivation.

Summary

- *Oxygen levels regulate the activity of numerous ion channels, strongly influencing their physiological activity.*
- *Regulation of ion channels by oxygen can account for the effects of hypoxia in specific tissues, particularly chemoreceptors.*
- *A number of different mechanisms have been put forward to account for hypoxic regulation of ion channels, but opinions remain divided as to their relative importance.*
- *The structural requirements of ion channels for oxygen-sensing has not been studied in depth, and represents an important area for future research.*

References

1. Lopez-Barneo, J., Pardal, R. & Ortega-Saenz, P. (2001) Cellular mechanism of oxygen sensing. *Ann. Rev. Physiol.* **63**, 259–287
2. Lopez-Barneo, J., Lopez-Lopez, J.R., Urena, J. & Gonzalez, C. (1988) Chemotransduction in the carotid body: K^+ current modulated by PO_2 in type I chemoreceptor cells. *Science* **241**, 580–582
3. Kline, D.D., Buniel, M.C., Glazebrook, P., Peng, Y.J., Ramirez-Navarro, A., Prabhakar, N.R. & Kunze, D.L. (2005) Kv1.1 deletion augments the afferent hypoxic chemosensory pathway and respiration. *J. Neurosci.* **25**, 3389–3399
4. Conforti, L., Bodi, I., Nisbet, J.W. & Millhorn, D.E. (2000) O_2-sensitive K^+ channels: role of the Kv1.2 subunit in mediating the hypoxic response. *J. Physiol.* **524**, 783–793
5. Hulme, J.T., Coppock, E.A., Felipe, A. Martens, J.R. & Tamkun, M.M. (1999) Oxygen sensitivity of cloned voltage-gated K^+ channels expressed in the pulmonary vasculature. *Circ. Res.* **85**, 489–497
6. Szigligeti, P., Neumeier, L., Duke, E., Chougnet, C., Takimoto, K., Lee, S.M., Filipovich, A.H. & Conforti, L. (2006) Signalling during hypoxia in human T lymphocytes-critical role of the src protein tyrosine kinase p56Lck in the O_2 sensitivity of Kv1.3 channels. *J. Physiol.* **573**, 357–370
7. Platoshyn, O., Brevnova, E.E., Burg, E.D., Yu, Y., Remillard, C.V. & Yuan, J.X. (2006) Acute hypoxia selectively inhibits KCNA5 channels in pulmonary artery smooth muscle cells. *Am. J. Physiol. Cell Physiol.* **290**, C907–C916
8. Finol-Urdaneta, R.K., Struver, N. & Terlau, H. (2006) Molecular and functional differences between heart mKv1.7 channel isoforms. *J. Gen. Physiol.* **128**, 133–145
9. Archer, S.L., Wu, X.C., Thebaud, B., Nsair, A., Bonnet, S., Tyrrell, B., McMurtry, M.S., Hashimoto, K., Harry, G. & Michelakis, E.D. (2004) Preferential expression and function of voltage-gated, O_2-sensitive K^+ channels in resistance pulmonary arteries explains regional heterogeneity in hypoxic pulmonary vasoconstriction: ionic diversity in smooth muscle cells. *Circ. Res.* **95**, 308–318
10. Patel, A.J., Lazdunski, M. & Honore, E. (1997) Kv2.1/Kv9.3, a novel ATP-dependent delayed-rectifier K^+ channel in oxygen-sensitive pulmonary artery myocytes. *EMBO J.* **16**, 6615–6625
11. Coppock E.A. & Tamkun, M.M. (2001) Differential expression of Kv channel α- and β-subunits in the bovine pulmonary arterial circulation. *Am. J. Physiol. Lung Cell Mol. Physiol.* **281**, L1350–L1360
12. Osipenko, O.N., Tate, R.J. & Gurney, A.M. (2000) Potential role for Kv3.1b channels as oxygen sensors. *Circ. Res.* **86**, 534–540
13. Wang, D., Youngson, C., Wong, V., Yeger, H., Dinauer, M.C., Vega-Saenz de Miera, E., Rudy, B. & Cutz, E. (1996) NADPH-oxidase and hydrogen peroxide sensitive K^+ channel may function as an oxygen sensor complex in airway chemoreceptors and small cell lung carcinoma cell lines. *Proc. Natl. Acad. Sci. U.S.A.* **93**, 13182–13187
14. Fu, X.W., Wang, D., Nurse, C.A., Dinauer, M.C. & Cutz, E. (2000) NADPH oxidase is an O_2 sensor in airway chemoreceptors: evidence from K^+ current modulation in wild type and oxidase-deficient mice. *Proc. Natl. Acad. Sci. U.S.A* **97,** 4374–4379
15. Perez-Garcia, M.T., Colinas, O., Miguel-Velado, E., Moreno-Dominguez, A. & Lopez-Lopez, J.R. (2004) Characterization of the Kv channels of mouse carotid body chemoreceptor cells and their role in oxygen sensing. *J. Physiol.* **557**, 457–471
16. Sanchez, D., Lopez-Lopez, J.R., Perez-Garcia, M.T., Sanz-Alfayate, G., Obeso, A., Ganfornina, M.D. & Gonzalez, C. (2002) Molecular identification of Kvα subunits that contribute to the oxygen-sensitive K^+ current of chemoreceptor cells of the rabbit carotid body. *J. Physiol.* **542**, 369–382
17. Li, Y.L. & Schultz, H.D. (2006) Enhanced sensitivity of Kv channels to hypoxia in the rabbit carotid body in heart failure: role of angiotensin II. *J. Physiol.* **575**, 215–227
18. Perez-Garcia, M.T., Lopez-Lopez, J.R. & Gonzalez, C. (1999) Kvβ1.2 subunit coexpression in HEK293 cells confers O_2 sensitivity to kv4.2 but not to Shaker channels. *J. Gen. Physiol.* **113**, 897–907

19. Lewis, A., Hartness, M.E., Chapman, C.G., Fearon, I.M., Meadows, H.J., Peers, C. & Kemp, P.J. (2001) Recombinant hTASK1 is an O_2-sensitive K^+ channel. *Biochem. Biophys. Res. Commun.* **285**, 1290–1294
20. Buckler, K.J., Williams, B.A. & Honore, E. (2000) An oxygen-, acid- and anaesthetic-sensitive TASK-like background potassium channel in rat arterial chemoreceptor cells. *J. Physiol.* **525**, 135–142
21. Hartness, M.E., Lewis, A., Searle, G.J., O'Kelly, I., Peers, C. & Kemp, P.J. (2001) Combined antisense and pharmacological approaches implicate hTASK as an airway O_2 sensing K^+ channel. *J. Biol. Chem.* **276**, 26499–26508
22. Johnson, R.P., O'Kelly, I.M. & Fearon, I.M. (2004) System specific O_2 sensitivity of the tandem pore domain K^+ channel TASK-1. *Am. J. Physiol. Cell Physiol.* **286**, C391–C397
23. O'Kelly, I., Lewis, A., Peers, C. & Kemp, P.J. (2000) O_2 sensing by airway chemoreceptor-derived cells: protein kinase C activation reveals functional evidence for involvement of NADPH oxidase. *J. Biol. Chem.* **275**, 7684–7692
24. Gurney, A.M., Osipenko, O.N., MacMillan, D., McFarlane, K.M., Tate, R.J. & Kempsill, F.E. (2003) Two-pore domain K channel, TASK-1, in pulmonary artery smooth muscle cells. *Circ. Res.* **93**, 957–964
25. Plant, L.D., Kemp, P.J., Peers, C., Henderson, Z. & Pearson, H.A. (2002) Hypoxic depolarization of cerebellar granule neurons by specific inhibition of TASK-1. *Stroke* **33**, 2324–2328
26. Caley, A.J., Gruss, M. & Franks, N.P. (2005) The effects of hypoxia on the modulation of human TREK-1 potassium channels. *J. Physiol.* **562**, 205–212
27. Miller, P., Kemp, P.J., Lewis, A., Chapman, C.G., Meadows, H.J. & Peers, C. (2003) Acute hypoxia occludes hTREK-1 modulation: re-evaluation of the potential role of tandem P domain K^+ channels in central neuroprotection. *J. Physiol.* **548**, 31–37
28. Miller, P., Kemp, P.J. & Peers, C. (2005) Structural requirements for O_2 sensing by the human tandem-P domain channel, hTREK1. *Biochem. Biophys. Res. Commun.* **331**, 1253–1256
29. Campanucci, V.A., Brown, S.T., Hudasek, K., O'Kelly, I.M., Nurse, C.A. & Fearon, I.M. (2005) O_2 sensing by recombinant TWIK-related halothane-inhibitable K^+ channel-1 background K^+ channels heterologously expressed in human embryonic kidney cells. Neuroscience **135**, 1087–1094
30. Campanucci, V.A., Fearon, I.M. & Nurse, C.A. (2003) A novel O_2-sensing mechanism in rat glossopharyngeal neurones mediated by a halothane-inhibitable background K^+ conductance. *J. Physiol.* **548**, 731–743
31. Peers,C. (1990) Hypoxic suppression of K^+ currents in type-I carotid-body cells - selective effect on the Ca^{2+}-activated K^+ current. *Neurosci. Lett.* **119**, 253–256
32. Riesco-Fagundo, A.M., Perez-Garcia, M.T., Gonzalez, C. & Lopez-Lopez, J.R. (2001) O_2 modulates large-conductance Ca^{2+}-dependent K^+ channels of rat chemoreceptor cells by a membrane-restricted and CO-sensitive mechanism. *Circ. Res.* **89**, 430–436
33. Lewis, A., Peers, C., Ashford, M.L. & Kemp, P.J. (2002) Hypoxia inhibits human recombinant large conductance, Ca^{2+}-activated K^+ (maxi-K) channels by a mechanism which is membrane delimited and Ca^{2+} sensitive. *J. Physiol.* **540**, 771–780
34. Keating, D.J., Rychkov, G.Y. & Roberts, M.L. (2001) Oxygen sensitivity in the sheep adrenal medulla: role of SK channels. *Am. J. Physiol. Cell Physiol.* **281**, C1434–C1441
35. Park, W.S., Han, J., Kim, N., Ko, J.H., Kim, S.J. & Earm, Y.E. (2005) Activation of inward rectifier K^+ channels by hypoxia in rabbit coronary arterial smooth muscle cells. *Am. J. Physiol. Heart Circ. Physiol.* **289**, H2461–H2467
36. Kemp, P.J. (2006) Detecting acute changes in oxygen: will the real sensor please stand up? *Exp. Physiol.* **91**, 829–834
37. Michelakis, E.D., Thebaud, B., Weir, E.K. & Archer, S.L. (2004) Hypoxic pulmonary vasoconstriction: redox regulation of O_2-sensitive K^+ channels by a mitochondrial O_2-sensor in resistance artery smooth muscle cells. *J. Mol. Cell Cardiol.* **37**, 1119–1136
38. Guzy R.D. & Schumacker, P.T. (2006) Oxygen sensing by mitochondria at complex III: the paradox of increased reactive oxygen species during hypoxia. *Exp. Physiol.* **91**, 807–819

39. Evans, A.M., Mustard, K.J., Wyatt, C.N., Peers, C., Dipp, M., Kumar, P., Kinnear, N.P. & Hardie, D.G. (2005) Does AMP-activated protein kinase couple inhibition of mitochondrial oxidative phosphorylation by hypoxia to calcium signaling in O_2-sensing cells? *J. Biol. Chem.* **280**, 41504–41511
40. Williams, S.E., Wootton, P., Mason, H.S., Bould, J., Iles, D.E., Riccardi, D., Peers, C. & Kemp, P.J. (2004) Hemoxygenase-2 is an oxygen sensor for a calcium-sensitive potassium channel. *Science* **306**, 2093–2097
41. Jaggar, J.H., Leffler, C.W., Cheranov, S.Y., Tcheranova, D. E S. & Cheng, X. (2002) Carbon monoxide dilates cerebral arterioles by enhancing the coupling of Ca^{2+} sparks to Ca^{2+}-activated K^+ channels. *Circ. Res.* **91**, 610–617
42. Prabhakar, N.R., Dinerman, J.L., Agani, F.H. & Snyder, S.H. (1995) Carbon monoxide: a role in carotid body chemoreception. *Proc. Natl. Acac. Sci. U.S.A.* **92**, 1994–1997
43. Adachi, T., Ishikawa, K., Hida, W., Matsumoto, H., Masuda, T., Date, F., Ogawa, K., Takeda, K., Furuyama, K., Zhang, Y. et al. (2004) Hypoxemia and blunted hypoxic ventilatory responses in mice lacking heme oxygenase-2. *Biochem. Biophys. Res. Commun.* **320**, 514–522
44. Ortega-Saenz, P., Pascual, A., Gomez-Diaz, R. & Lopez-Barneo, J. (2006) Acute oxygen sensing in heme oxygenase-2 null mice. *J. Gen. Physiol.* **128**, 405–411
45. Balbir, A., Lee, H., Okumura, M., Biswal, S., Fitzgerald, R.S. & Shirahata, M. (2006) A search for genes that may confer divergent morphology and function in the carotid body between two strains of mice. *Am. J. Physiol. Lung Cell Mol. Physiol.* **292**, L704–L715
46. Franco-Obregon, A., Urena, J. & Lopez-Barneo, J. (1995) Oxygen-sensitive calcium channels in vascular smooth muscle and their possible role in hypoxic arterial relaxation, *Proc. Natl. Acad. Sci. U.S.A.* **92**, 4715–4719
47. Franco-Obregon, A. & Lopez-Barneo, J. (1996) Low PO_2 inhibits calcium channel activity in arterial smooth muscle cells. *Am. J. Physiol.* **271**, H2290–H2299
48. Franco-Obregon, A. & Lopez-Barneo, J. (1996) Differential oxygen sensitivity of calcium channels in rabbit smooth muscle cells of conduit and resistance pulmonary arteries [published erratum appears in J. Physiol. (Lond) 1996 Jun 15;493(Pt 3):923], *J. Physiol.* **491**, 511–518
49. Fearon, I.M., Palmer, A.C.V., Balmforth, A.J., Ball, S.G., Mikala, G., Schwartz, A. & Peers, C. (1997) Hypoxia inhibits the recombinant α1C subunit of the human cardiac L-type Ca^{2+} channel. *J. Physiol.* **500**, 551–556
50. Fearon, I.M., Varadi, G., Koch, S., Isaacsohn, I., Ball, S.G. & Peers, C. (2000) Splice variants reveal the region involved in oxygen sensing by recombinant human L-type Ca^{2+} channels. *Circ. Res.* **87**, 537–539
51. Conforti, L., Takimoto, K., Petrovic, M., Pongs, O. & Millhorn, D. (2003) The pore region of the Kv1.2α subunit is an important component of recombinant Kv1.2 channel oxygen sensitivity. *Biochem. Biophys. Res. Commun.* **306**, 450–456

7

Cellular mechanisms associated with intermittent hypoxia

Jayasri Nanduri and Nanduri R. Prabhakar[1]

Center for Systems Biology, Department of Medicine, The University of Chicago, Chicago, IL 60637, U.S.A.

Abstract

Hypoxia, i.e. decreased availability of oxygen occurs under many different circumstances and can be either continuous or intermittent. Continuous hypoxia such as that experienced during periods of high altitude leads to physiological adaptations, whereas chronic IH (intermittent hypoxia) associated with sleep-disordered breathing manifested as recurrent apneas leads to morbidity. The purpose of the present chapter is to highlight recent findings on cellular responses to IH. Studies on cell culture models of IH revealed that for a given duration and intensity, IH is more potent than continuous hypoxia in evoking transcriptional activation. IH activates HIF-1 (hypoxia-inducible factor-1), the immediate early gene c-fos, activator protein-1, nuclear factor κB and cAMP-response-element-binding protein. Physiological studies showed that HIF-1 plays an important role in chronic IH-induced autonomic abnormalities in mice. IH affects expression of proteins associated with neuronal survival and apoptosis, as well as post-translational modifications of proteins resulting in increased biological activity. Comparisons between continuous hypoxia and IH revealed notable differences in the kinetics of protein kinase activation, type of protein kinase being activated and the downstream targets of protein kinases. IH increases

[1]*To whom correspondence should be addressed (email nanduri@uchicago.edu).*

ROS (reactive oxygen species) generation both in cell culture and in intact animals, and ROS-mediated signalling mechanisms contribute to cellular and systemic responses to IH. Future studies utilizing genomic and proteomic approaches may provide important clues to the mechanisms by which IH leads to morbidity as opposed to continuous hypoxia-induced adaptations. Cellular mechanisms associated with IH (other than recurrent apneas) such as repetitive, brief ascents to altitude, however, remain to be studied.

Introduction

Molecular oxygen is essential for the survival of mammalian cells because of its critical role in generating ATP via oxidative phosphorylation. Hypoxia, i.e. the decreased availability of oxygen, occurs under many different circumstances. Continuous hypoxia is experienced during periods at high altitude. Physiological systems adapt to continuous hypoxia, and the biochemical and molecular mechanisms underpinning the adaptations have been extensively investigated [1,2].

IH (intermittent hypoxia) is more often experienced than continuous hypoxia in people living at sea level. Chronic IH is experienced as a consequence of sleep-disordered breathing (manifested as recurrent sleep apneas) in 50% of premature infants [3], 5% of middle-aged men and 2% of women after menopause [4,5]. In this condition, transient repetitive apnea (cessation of breathing) results in periodic hypoxaemia [decreases in arterial blood PO_2 (partial pressure of oxygen)]. Each episode of apnea lasts no more than tens of seconds and the frequency of apneas may exceed 60 episodes per hour with blood haemoglobin saturation of oxygen reduced to as low as 50% in severely affected patients. Recurrent apnea patients in addition to chronic IH also experience chronic intermittent hypercapnia [elevations in arterial blood PCO_2 (partial pressure of carbon dioxide)]. A major advance in the field of apnea research is the demonstration that exposing experimental animals to chronic IH alone is sufficient to result in physiological changes similar to those described in recurrent apnea patients [6]. Unlike continuous hypoxia, humans experiencing chronic IH as a consequence of recurrent apneas exhibit morbidity including development of hypertension, myocardial infarctions and stroke [4,5]. The purpose of this chapter is to summarize recent information on cellular responses to the paradigm that IH simulates recurrent apneas and compare these with the responses to continuous hypoxia.

Transcriptional activation during IH

It is being increasingly recognized that activation of specific genes is an important mechanism by which hypoxia triggers long-term adaptive responses [2]. The following section describes the transcriptional changes associated with IH and highlights differences between the effects of continuous hypoxia and IH.

HIF-1 (hypoxia inducible factor-1)

The transcriptional activator HIF-1 is a global regulator of oxygen homoeostasis that controls multiple key developmental and physiological processes [7]. Over 60 HIF-1 target genes have been identified, including those encoding EPO (erythropoietin) and VEGF (vascular endothelial growth factor) [8]. HIF-1 is a heterodimeric protein that is composed of a constitutively expressed HIF-1β subunit and an oxygen-regulated HIF-1α subunit. HIF-1 activity is induced under conditions of continuous hypoxia as a result of a decreased rate of oxygen-dependent proline hydroxylation, ubiquitination and proteasomal degradation of the HIF-1α subunit [9]. HIF-1 transcriptional activity is also regulated via oxygen-dependent asparginine hydroxylation that blocks co-activator recruitment [10].

HIF-1 activation by IH was studied in a cell culture model of IH wherein PC12 (pheochromocytoma-12) cells were exposed to alternating cycles of hypoxia (1.5% oxygen for 30 s) and re-oxygenation (20% oxygen for 4 min) [11]. Medium PO_2 decreased from 74±3 mmHg to 55±5 mmHg during each episode of hypoxia. Reintroduction of normoxic gas restored the medium PO_2 to the pre-hypoxic value. Cell viability, pH and osmolarity of the medium were unaffected up to 120 cycles of IH [11]. Under these experimental conditions, HIF-1α protein increased in a stimulus-dependent manner as the duration of IH increased from 10 to 30 to 60 cycles [12] To study HIF-1 transcriptional activity, PC12 cells were transfected with a reporter gene in which the expression of firefly luciferase was driven by a HIF-1-dependent HRE (hypoxia-response element) upstream of an SV40 promoter [12]. IH resulted in significant stimulus-dependent HRE transcriptional activity. These observations suggest that IH, despite a modest decrease in medium PO_2 (~20–25 mmHg) was effective in inducing HIF-1α protein expression and HIF-1 transcriptional activity, two necessary pre-requisites for HIF-1-dependent gene transcription.

On the other hand, when cells were exposed to 60 min of continuous hypoxia, which was equivalent to the hypoxic duration accumulated during 120 episodes of IH (30 s each episode), the medium PO_2 decreased from 80±6 to 10±2 mmHg (a decrease of 70 mmHg) within 5 min after the onset of hypoxia and remained at this level during the entire period of hypoxic exposure. Despite the marked reduction in PO_2, 60 min of continous hypoxia was ineffective in increasing either HIF-1α protein expression or activation of HRE [12]. However, extending the hypoxic duration to 4 h did result in a significant increase in HIF-1α protein expression as well as HRE activation [12], as reported by others [13]. These observations suggest that for a given duration and intensity, IH is more potent in activating HIF-1 than continuous hypoxia.

Mechanisms associated with IH-induced HIF-1α stabilization and activation

A recent brief communication by Yuan et al. [14] suggests that HIF-1α stabilization by IH involves increased protein synthesis and activation of rapamycin-sensitive mTOR (mammalian target of rapamycin) signalling. These investigators reported that rapamycin inhibits IH-induced HIF-1α stabilization and IH increases phosphorylated mTOR levels as well as downstream S6 kinases. Furthermore, the effects of IH on mTOR activation as well as on HIF-1α protein can be prevented by inhibitors of IP_3 (inositol-3 phosphate) receptors and PLC-γ (phospholipase C-γ) as well as calcium chelators, indicating that mobilization of intracellular calcium is an important upstream signalling event required for IH-induced mTOR activation and subsequent HIF-1α accumulation. These observations are intriguing in that they are in striking contrast to the inhibition of the mTOR signalling pathway reported with continuous hypoxia [15]. However, Hui et al. [16] reported that during continuous hypoxia, translation of HIF-1α can be stimulated by an influx of extracellular calcium and activation of the classical PKC (protein kinase C) and mTOR pathways in PC12 cells. Whether proline hydroxylation and subsequent ubiquination pathways, which are critical for HIF-1α stabilization during continuous hypoxia, also play a role during IH requires further study.

HIF-1 activation by continuous hypoxia requires activation of PI3K (phosphoinositide 3-kinase) [17]. However, neither LY294002 nor wortmannin, inhibitors of PI3K, were able to block IH-induced HIF-1 transcriptional activity. On the other hand, the calcium chelator BAPTA-AM prevented IH-induced HRE activation, indicating the involvement of calcium signalling pathways in IH-induced HIF-1 transcriptional activity [12]. CaMKs (calcium/calmodulin-dependent protein kinases) are downstream signalling molecules that participate in calcium-mediated gene regulation. A previous study reported that continuous hypoxia leads to a transient (15 min) and modest (1.5-fold) increase in CaMKII activity in PC12 cells [18]. In striking contrast, IH resulted in an exponential and nearly 6-fold increase in CaMKII activity in response to increasing cycles of IH and was associated with increased phosphorylation of CaMKII protein [12]. Co-transfection of constitutively active CaMKII-290 mimicked the effects of IH. Inhibitors of either calmodulin (W-7) or CaMK (KN-93) prevented IH-induced but not continuous hypoxia-induced HIF-1 transcriptional activity [12]. A CaMK inhibitor, however, was ineffective in preventing IH-induced HIF-1α protein expression, suggesting that CaMK-dependent signalling is critical for IH-induced HIF-1 transcriptional activation but not for HIF-1α protein expression. IH stimulated the CAD (C-terminal transactivation domain) of HIF-1α via a mechanism that is independent of Asn^{830} hydroxylation and CaMKII activation was required for this response [12]. On the other hand, Asn^{830} hydroxylation is required for continuous hypoxia-evoked CAD activation [10]. How might CaMKII contribute to IH-induced HIF-1 transactivation? Several lines of evidence suggest that p300/CBP proteins are the major co-activators for IH-induced HIF-1 activation [19,20]. IH increased p300 transcriptional activity, which was blocked by KN-93 [12]. The signalling pathways associated with HIF-1 activation by IH are schematically presented in Figure 1. These observations suggest that IH-induced HIF-1 transactivation requires a novel signalling pathway involving

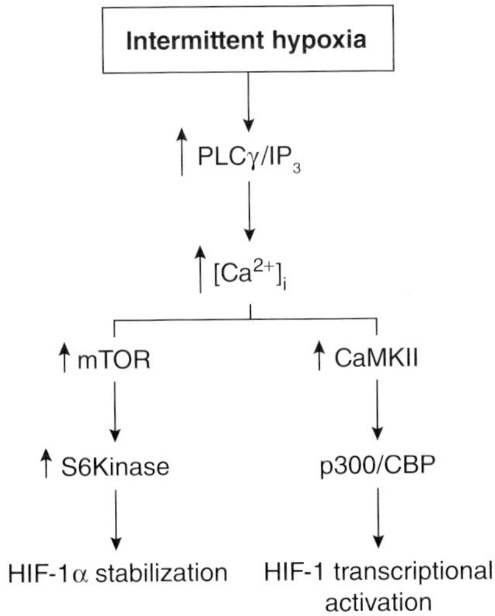

Figure 1. Cellular mechanisms associated with activation of HIF-1 by IH

CaMKII-dependent activation of p300 co-activator, suggesting striking differences in the signalling pathways associated with HIF-1 activation in response to IH compared with continuous hypoxia.

Functional significance of HIF-1 activation by IH

Patients with IH caused by recurrent apneas develop autonomic abnormalities including systemic hypertension, heightened ventilatory response to hypoxia and persistent activation of the sympathetic nervous system [21]. IH-induced autonomic abnormalities have been attributed to activation of the carotid body, the primary sensory organ for detecting systemic hypoxia [21]. Previous studies have shown that rodents exposed to IH also develop systemic hypertension, elevated plasma catecholamines, augmented hypoxic ventilatory response and elevated sympathetic nerve activity [22] similar to that reported in recurrent apnea patients. Supporting a role for the carotid body, it has been shown that chronic IH sensitizes chemoreceptor response to acute hypoxia and induces long-lasting sensory excitation [sensory LTF (long-term facilitation)] in rats [23]. Alterations in the carotid body activity may contribute to increased sympathetic tone and systemic hypertension.

Complete HIF-1α deficiency results in embryonic lethality at mid-gestation, whereas hif-1α$^{+/-}$ HET (heterozygous) mice, which are partially deficient in HIF-1α expression, develop normally and are indistinguishable from WT (wild-type) littermates under normoxic conditions [24]. Using hif-1α$^{+/-}$ HET mice, a recent study examined the importance of HIF-1 activation in IH-induced physiological alterations [25]. In contrast to WT mice, IH was ineffective in evoking HIF-1α protein expression in HET mice [25]. More importantly, the absence of a HIF-1α response to IH in HET mice was associated with a striking absence of blood pressure elevation, plasma catecholamines and augmented ventilatory response to hypoxia as well as an absence of carotid body sensitization and sensory LTF [25].

Clinical studies indicate that recurrent apneas are associated with increased serum levels of EPO and VEGF [26,27], two target genes activated by HIF-1. The effects of IH-induced elevations in circulating EPO on cardiac responses to ischaemia–reperfusion have been examined [28,29]. Exogenous EPO prevented deleterious effects of ischaemia–reperfusion in the heart [28]. Studies in WT mice have shown that EPO expression was elevated by IH with a concomitant cardiac protection against ischaemia–reperfusion injury [29]. In contrast, such cardiac protection against ischaemia–reperfusion was absent in IH-exposed hif-1$\alpha^{+/-}$ mice, demonstrating a critical involvement of HIF-1 in these responses [29]. These observations, taken together with the biochemical data, suggest that IH activates HIF-1-mediated transcription which appears to be critical for evoking cardio-respiratory responses. However, future studies are needed to identify the HIF-1-regulated downstream genes other than EPO and VEGF that contribute to IH-evoked cardio-respiratory changes.

Immediate early genes

Hypoxia is a potent activator of the *c-fos* gene, a member of the immediate early gene family [30]. c-Fos protein in combination with c-Jun, another member of the immediate early gene family, forms the heterodimeric AP-1 (activator protein-1) complex. AP-1 functions as a transcriptional activator that drives expression of a variety of genes. Greenberg et al. [31] reported up-regulation of c-Fos protein in the central nervous system of rats exposed to 30 days of IH. Yuan et al. [11] examined the effects of IH on c-fos mRNA expression in PC12 cells. They reported that IH increases *c-fos* mRNA expression in a stimulus-dependent manner. The IH-induced increase in *c-fos* mRNA was in part due to increased *c-fos* transcriptional activation as evidenced by a reporter gene assay. Point mutations in the *c-fos* promoter showed that SRE (serum responsive element), CRE (calcium response element) but not FAP [AP-1/CRE (cAMP-response-element)-like cis-element] are critical for IH-induced *c-fos* promoter activation [11]. Interestingly, the magnitude of *c-fos* activation by IH was dependent on the duration of re-oxygenation between the hypoxic episodes, and not on the duration of the hypoxic episodes [11]. Exposure to continous hypoxia (comparable cumulative duration of hypoxia accumulated during IH) had virtually no effect on *c-fos* expression [11]. These observations reiterate that for a given duration, IH is a more potent transcriptional activator than continuous hypoxia. Interestingly, following termination of IH *c-fos* mRNA remained elevated for at least 3 h, whereas it returned to control levels within 1 h of terminating continuous hypoxia [11], demonstrating another striking difference between both forms of hypoxia. The long-lasting *c-fos* mRNA activation by IH is reminiscent of IH-induced long-lasting activation of the carotid body sensory activity [23] and breathing [32,33] reported in intact animals. These findings indicate that long-lasting activation, which is a hallmark of IH in intact animals, can be elicited by IH even at the gene level.

The mechanisms by which IH leads to long-lasting gene activation, however, remain to be studied.

What might be the functional significance of IH-evoked *c-fos* activation? IH resulted in increased AP-1 transcriptional activity and expression of mRNA of TH (tyrosine hydroxylase) a downstream AP-1-regulated gene [11]. Co-transfection with antisense *c-fos* abolished IH-induced AP-1 activation and TH mRNA expression, implying that *c-fos* contributes to IH-induced AP-1 activation and downstream gene transcription [11]. Since TH is the rate-limiting enzyme in catecholamine synthesis, it is likely that IH-induced TH activation contributes in part to the increased catecholamine levels reported in IH-exposed animals [34], as well as in humans experiencing IH due to apneas [35].

NF-κB (nuclear factor κB)

NF-κB is an important transcriptional regulator of inflammatory mediators. Ryan et al. [36] reported that in obstructive sleep apnea patients NF-κB activity in monocytes is increased resulting in elevated serum TNFα (tumour necrosis factor-α) and IL-8 (interleukin-8) levels. Elevated TNFα levels correlated with arterial oxygen desaturation, and treating apnea patients with CPAP (continuous positive airway pressure) normalized TNFα levels [36]. These observations led to the suggestion that IH activates NF-κB. Consistent with this possibility is the demonstration that IH stimulates NF-κB-mediated transcriptional activation in HeLa cells [37]. In addition, Greenberg et al. [38] found elevated NF-κB activity in IH-exposed mice and apnea patients. The mechanisms by which IH activates NF-κB, however, remain to be investigated.

The studies described thus far suggest that IH is a potent activator of HIF-1, *c-fos* and NF-κB. Whether IH activates other transcriptional factors requires further study.

Altered protein expression

The following section summarizes the effects of IH on protein expression and post-translational modifications.

Protein expression leading to neuronal injury and apoptosis

The influence of IH on acid-base transporters in the mouse central nervous system was investigated by Douglas et al. [39]. In this study, 2- to 3-day-old mice were exposed to alternating cycles of 2 min of hypoxia (6.0–7.5% oxygen) and 3 min of normoxia (21% oxygen) for 8 h/day for 28 days. IH exposure decreased the expression of sodium/hydrogen exchangers (the NHE-1 and 3 isoforms) and sodium bicarbonate co-transporter proteins in the central nervous system, especially in the cerebellum. It has been proposed that IH, by down-regulating the acid-extruding capacity of neurons, renders them more susceptible to acidic insult and subsequent neuronal injury.

Recent studies suggest that IH leads to cognitive deficits and is associated with neuronal apoptosis in experimental animals [40]. Acute exposure to IH (10% oxygen followed by 21% oxygen for 90 s each) for 6 h resulted in a much greater injury of the CA1 region of the hippocampus than the CA3 region. Proteomic analysis of the CA1 and CA3 regions of the hippocampus derived from IH and control rats demonstrated a complex pattern of protein alterations that accompanied CA1 injury [41]. IH augmented the expression of stress-induced proteins, primarily chaperone proteins and proteins related to apoptosis. In addition, comparison of protein maps of the CA1 and CA3 regions revealed a marked up-regulation of a hitherto uncharacterized protein in the CA1 region but not in the CA3 region, which might play a key role in CA1 injury during IH [42].

Exposure of rats to alternating cycles of hypoxia (10% oxygen for 90 s) and normoxia (21% oxygen for 90 s) for up to 14 days has been shown to selectively up-regulate protein expression and activity of COX (cyclo-oxygenase)-2, but not COX-1, in cortical regions of the brain [43]. The up-regulation of COX-2 was associated with increased neuronal apoptosis and neurobehavioural abnormalities as evidenced by deficits in the acquisition and retention of a spatial task [43]. Furthermore, studies by Goldbart et al. [44] suggest that decreased phosphorylation of CREB (cAMP-response-element-binding protein) accounts for decreased neuronal survival and spatial learning defects observed in IH exposed rats. Gozal et al. [45] showed that in PC12 cells, IH induced an earlier and more extensive apoptotic response than continuous hypoxia that was associated with enhanced caspase activity.

Post-translational modifications of proteins

Hypoxia, in addition to affecting protein expression, also alters the function of existing proteins via post-translational modifications of specific amino acid residues involving phosphorylation and dephosphorylation reactions. Studies on PC12 cells showed that IH activates a variety of protein kinases including MAPKs [mitogen-activated protein kinases; ERK1/2 (extracellular-signal-regulated kinase 1/2) and Jun kinase], PKC and CaMKII [12]. Comparison of continuous hypoxia with IH revealed some notable differences in protein kinase activation. For instance, IH resulted in a sustained increase in CaMKII (5.5-fold) [12], whereas continuous hypoxia caused a transient (15 min) and modest increase (1.5-fold activation) [18]. Likewise, Jun kinase was only activated by IH but not by continuous hypoxia [12]. Continuous hypoxia but not IH activates p38 kinase in PC12 cells [12,18]. Although both IH and continuous hypoxia activate ERK-1/2 in PC12 cells, ERK-mediated signalling is critical for HIF-1 activation by continuous, but not by IH [12]. Thus, although both IH and continuous hypoxia activates similar protein kinases, the kinetics of activation and the downstream targets seem to differ between the two forms of hypoxia.

In PC12 cells IH increased the activity of the TH enzyme, which is in part due to increased phosphorylation of the Ser^{40} residue [46]. IH-induced TH activation could be prevented by inhibitors of CaMK and PKA (protein kinase A) [46]. Gozal et al. [47] compared the changes in TH mRNA, protein expression and activity in various brain regions of male rats exposed to continuous hypoxia and IH, and concluded that differences in TH activity between IH and continuous hypoxia are related to different phosphorylation patterns. Studies by Goldbart et al. [44] showed that IH results in a time-dependent decrease in phosphorylation and nuclear binding of CREB without changes in total CREB protein. It will be of interest to determine whether other types of post-translational events like sulfation, glycosylation and ubiquitination also play a role in altered protein activities associated with IH.

ROS (reactive oxygen species) signalling during IH

It is evident from studies outlined above that IH and continuous hypoxia have profoundly different effects on transcriptional activation, signalling pathways and protein expression. What makes IH different from continuous hypoxia? A major difference is that IH is interspersed with periods of re-oxygenation, which are absent during continuous hypoxia. In cell cultures the duration of re-oxygenation rather than the hypoxic phase was found to influence IH-induced gene expression [11]. Thus it was proposed that IH might generate ROS during the re-oxygenation phase analogous to that reported with ischaemia–reperfusion.

ROS generation during IH

Aconitase is an enzyme in the tricarboxylic acid cycle and is present both in the cytosol and in mitochondria. ROS inhibits aconitase activity and antioxidants prevent this response. Thus inhibition of aconitase activity can serve as an index of ROS generation [48]. Aconitase activity was markedly down regulated in the carotid body [23] and in adrenal medulla [34] from IH-exposed rats as well as in IH exposed PC12 cells [11]. Likewise, TBARS (thiobarbituric acid reactive substances), an indicator of protein oxidation are elevated in tissues from IH-exposed animals and antioxidants prevented this response [25]. Taken together, these observations suggest that IH increases ROS generation. However, these studies do not indicate which of the reactive oxygen species (i.e. O_2^-, H_2O_2 or OH^-) are elevated in response to IH.

ROS generation during IH could involve inhibition of the mitochondrial electron transport chain complex as well as activation of several oxidases. IH decreases mitochondrial complex I but not complex III activity in cell cultures [11]. On the other hand, there is evidence suggesting that continuous hypoxia generates ROS by inhibiting complex III of the electron transport chain [49]. The mechanisms by which IH selectively targets complex I and the potential contribution of oxidases to IH-induced ROS generation remain to be investigated. MnTMPyP [manganese (III) tetrakis-(1-methyl-4-pyridyl)-porphyrin pentachloride], a superoxide dismutase mimetic that scavenges superoxide anion without generating H_2O_2, prevented up regulation of *c-fos* and AP-1

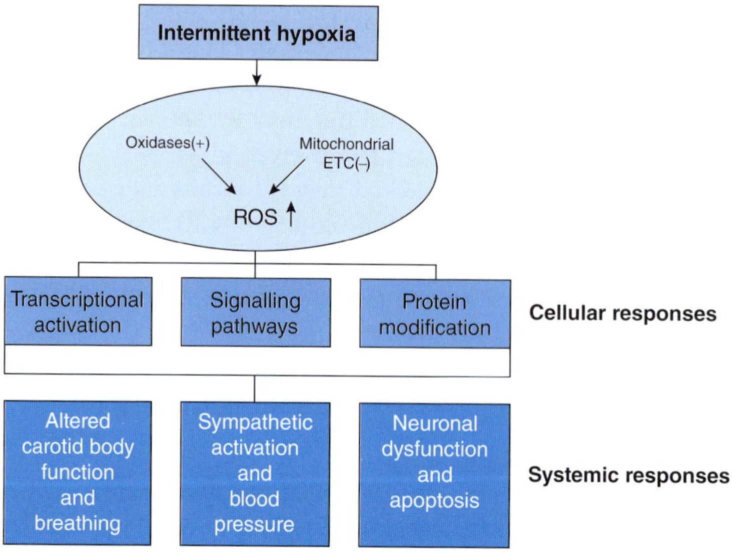

Figure 2. Schematic representation of the role of ROS in cellular and systemic responses to IH
ETC, electron transport chain.

activation in PC12 cell cultures exposed to IH [11]. Likewise, MnTMPyP prevented IH-induced HIF-1α accumulation [14]. Furthermore, the observation that MnTMPyP also prevents IH-induced activation of Jun kinase [50] suggests a role for ROS in IH-induced protein kinase activation. ROS might activate protein kinases either directly or indirectly by inhibiting protein phosphatases or a combination of both, clarification of which requires further study. Physiological studies have shown that MnTMPyP as well as N-acetyl cysteine, a precursor of glutathione (another antioxidant), prevent IH-induced elevations in blood pressure, plasma catecholamines [34] and augmented ventilatory response to hypoxia as well as functional alterations in carotid body activity [25]. These studies suggest that ROS signalling plays an important role in mediating biochemical and physiological responses to IH (Figure 2).

Conclusion

In summary, the studies described above suggest that IH, such as that associated with recurrent apneas, affects a variety of biochemical processes, including transcriptional activation of genes and signalling mechanisms involving post-translational modification of proteins that may underlie the physiological responses. Emerging evidence indicates that ROS-mediated signalling is critical for evoking IH-induced cellular and systemic responses. However, it must be pointed out that our understanding of the physiological and cellular responses to IH is in its very early stages. There are several gaps in our knowledge as to the mechanism by which IH affects gene and protein expression. Except for a few proteins, there is a paucity of information regarding whether hypoxia-induced up-regulation of proteins is due to *de novo* synthesis or to increased protein stabilization, or to translocation of proteins to specific cellular compartments. Thus far studies examining the influence of IH have focused on those proteins that show increased expression. But it is equally important to assess the role of proteins whose expression is down-regulated during IH. Future studies using genomic and proteomic approaches may provide important clues to the process by which IH leads to morbidity as opposed to the adaptations induced by continuous hypoxia. In addition to recurrent apneas, humans also experience IH during repetitive brief ascents to high altitude, wherein each hypoxic episode lasts for a few hours [51]. It remains to be determined whether both forms of IH activate similar cellular mechanisms.

Summary

- *Hypoxia (i.e. a decrease in arterial blood oxygen) can be continuous or intermittent.*
- *Continuous hypoxia such as that experienced during sojourns at high altitude leads to physiological adaptations, whereas chronic IH, associated with recurrent apneas, results in morbidity.*
- *For a given duration and intensity, IH is more potent than continuous hypoxia in activating various transcription factors including HIF-1 and c-fos.*
- *Physiological studies revealed that HIF-1-mediated transcription plays a critical role in mediating IH-induced cardio-respiratory responses.*
- *The kinetics of protein kinase activation and the type of kinase activated by IH differ considerably from continuous hypoxia.*
- *IH increases ROS generation and ROS-mediated signalling mechanisms play a critical role in IH-induced cellular and systemic responses.*

We are grateful for the grant support from the National Institutes of Health, Heart, Lung and Blood Institute (HL-25830 and HL-46462).

References

1. Bunn, H.F. & Poyton, R.O. (1996) Oxygen sensing and molecular adaptation to hypoxia. *Physiol. Rev.* **76**, 839–885
2. Semenza, G.L. (1999) Perspectives on oxygen sensing. *Cell* **6**, 281–284
3. Poets, C.F., Samuels, M.P. & Southall, D.P. (1994) Epidemiology and pathophysiology of apnoea of prematurity. *Biol. Neonate* **65**, 211–219
4. Nieto, F.J., Young, T.B., Lind, B.K., Shahar, E., Samet, J.M., Redline, S., D'Agostino, R.B., Newman, A.B., Lebowitz, M.D. & Pickering, T.G. (2000) Association of sleep-disordered breathing, sleep apnea, and hypertension in a large community-based study. Sleep Heart Health Study. *JAMA* **283**, 1829–1836
5. Shahar, E., Whitney, C.W., Redline, S., Lee, E.T., Newman, A.B., Javier Nieto, F., O'Connor, G.T., Boland, L.L., Schwartz, J.E. & Samet, J.M. (2001) Sleep-disordered breathing and cardiovascular

disease: cross-sectional results of the Sleep Heart Health Study. *Am. J. Respir. Crit. Care Med.* **163**, 19–25
6. Fletcher, E.C. (2001) Physiological consequences of intermittent hypoxia: systemic blood pressure. *J. Appl. Physiol.* **90**, 1600–1605
7. Semenza, G.L. (2000) HIF-1: mediator of physiological and pathophysiological responses to hypoxia. *J. Appl. Physiol.* **88,** 1474–1480
8. Semenza, G.L. (2000) Expression of hypoxia-inducible factor 1: mechanisms and consequences. *Biochem. Pharmacol.* **59**, 47–53
9. Hirota, K. & Semenza, G.L. (2005) Regulation of hypoxia-inducible factor 1 by prolyl and asparaginyl hydroxylases. *Biochem. Biophys. Res. Commun.* **338**, 610–616
10. Lando, D., Peet, D.J., Whelan, D.A., Gorman, J.J. & Whitelaw, M.L. (2002) Asparagine hydroxylation of the HIF transactivation domain a hypoxic switch. *Science* **295**, 858–861
11. Yuan, G., Adhikary, G., McCormick, A.A., Holcroft, J.J., Kumar, G.K. & Prabhakar, N.R. (2004) Role of oxidative stress in intermittent hypoxia-induced immediate early gene activation in rat PC12 cells. *J. Physiol.* **557**, 773–783
12. Yuan, G., Nanduri, J., Bhasker, C.R., Semenza, G.L. & Prabhakar, N.R. (2005) Ca^{2+}/calmodulin kinase-dependent activation of hypoxia inducible factor 1 transcriptional activity in cells subjected to intermittent hypoxia. *J. Biol. Chem.* **280**, 4321–4328
13. Iyer, N.V., Kotch, L.E., Agani, F., Leung, S.W., Laughner, E., Wenger, R.H., Gassman, M., Gearhart, J.D., Lawler, A.M., Yu, A.Y. & Semenza, G.L. (1998) Cellular and developmental control of O_2 homeostasis by hypoxia-inducible factor 1 α. *Genes Dev.* **12**, 149–162
14. Yuan, G., Nanduri, J., Semenza, G. & Prabhakar, N.R. (2006) Mechanisms of HIF-1 stabilization by intermittent hypoxia: role of Ca^{2+}-mTOR signaling. *FASEB J.* **20**, A790
15. Liu, L., Cash, T.P., Jones, R.G., Keith, B., Thompson, C.B. & Simon, M.C. (2006) Hypoxia-induced energy stress regulates mRNA translation and cell growth. *Mol. Cell* **21**, 521–531
16. Hui, A.S., Bauer, A.L., Striet, J.B., Schnell, P.O. & Czyzyk-Krzeska, M.F. (2006) Calcium signaling stimulates translation of HIF-α during hypoxia. *FASEB J.* **20**, 466–475
17. Zhou, J., Schmid, T., Frank, R. & Brune, B. (2004) PI3K/Akt is required for heat shock proteins to protect hypoxia-inducible factor 1α from pVHL-independent degradation. *J. Biol. Chem.* **279**, 13506–13513
18. Premkumar, D.R., Mishra, R.R., Overholt, J.L., Simonson, M.S., Cherniack, N.S. & Prabhakar, N.R. (2000) L-type Ca(2+) channel activation regulates induction of c-fos transcription by hypoxia. *J. Appl. Physiol.* **88**, 1898–1906
19. Dames, S.A., Martinez-Yamout, M., De Guzman, R.N., Dyson, H.J. & Wright, P.E. (2002) Structural basis for HIF-1α/CBP recognition in the cellular hypoxic response. *Proc. Natl. Acad. Sci. U.S.A.* **99**, 5271–5276
20. Sang, N., Fang, J., Srinivas, V., Leshchinsky, I. & Caro, J. (2002) Carboxyl-terminal transactivation activity of hypoxia-inducible factor 1α is governed by a von Hippel-Lindau protein-independent, hydroxylation-regulated association with p300/CBP. *Mol. Cell Biol.* **22**, 2984–2992
21. Prabhakar, N.R., Dick, T.E., Nanduri, J. & Kumar, G.K. (2007) Systemic cellular and molecular analysis of chemoreflex-mediated sympatho-excitation by chronic intermittent hypoxia. *Exp. Physiol.* **92**, 39–44
22. Prabhakar, N.R., Fields, R.D., Baker, T. & Fletcher, E.C. (2001) Intermittent hypoxia: cell to system. *Am. J. Physiol. Lung Cell Mol. Physiol.* **281**, L524–L528
23. Peng, Y.J., Overholt, J.L., Kline, D., Kumar, G.K. & Prabhakar, N.R. (2003) Induction of sensory long-term facilitation in the carotid body by intermittent hypoxia: implications for recurrent apneas. *Proc. Natl. Acad. Sci. U.S.A.* **100**, 10073–10078
24. Yu, A.Y., Shimoda, L.A., Iyer, N.V., Huso, D.L., Sun, X., McWilliams, R., Beaty, T., Sham, J.S., Wiener, C.M., Sylvester, J.T. & Semenza, G.L. (1999) Impaired physiological responses to chronic hypoxia in mice partially deficient for hypoxia-inducible factor 1α. *J. Clin. Invest.* **103**, 691–696
25. Peng, Y.J., Yuan, G., Ramakrishnan, D., Sharma, S.D., Bosch-Marce, M., Kumar, G.K., Semenza, G.L. & Prabhakar, N.R. (2006) Heterozygous HIF-1α deficiency impairs carotid body-mediated

systemic responses and reactive oxygen speices generation in mice exposed to intermittent hypoxia. *J. Physiol.* **577**, 705–716

26. Gozal, D., Lipton, A.J. & Jones, K.L. (2002) Circulating vascular endothelial growth factor levels in patients with obstructive sleep apnea. *Sleep* **25**, 59–65
27. Schulz, R., Hummel, C., Heinemann, S., Seeger, W. & Grimminger, F. (2002) Serum levels of vascular endothelial growth factor are elevated in patients with obstructive sleep apnea and severe nighttime hypoxia. *Am. J. Respir. Crit. Care Med.* **165**, 67–70
28. Ramond, A, Ribuot, C., Levy, P. & Joyeux-Faure, M. (2007) Deleterious myocardial consequences induced by intermittent hypoxia are reversed by erythropoietin. *Respir. Physiol. Neurobiol.* **156**, 362–369
29. Cai, Z., Manalo, D.J., Wei, G., Rodriguez, E.R., Fox-Talbot, K., Lu, H., Zweier, J.L. & Semenza, G.L. (2003) Hearts from rodents exposed to intermittent hypoxia or erythropoietin are protected against ischemia-reperfusion injury. *Circulation* **108**, 79–85
30. Premkumar, D.R., Adhikary, G., Overholt, J.L., Simonson, M.S., Cherniack, N.S. & Prabhakar, N.R. (2000) Intracellular pathways linking hypoxia to activation of c-fos and AP-1. *Adv. Exp. Med. Biol.* **475**,101–109
31. Greenberg, H.E., Sica, A.L., Scharf, S.M. & Ruggiero, D.A. (1999) Expression of c-fos in the rat brainstem after chronic intermittent hypoxia. *Brain Res.* **816**, 638–645
32. Peng, Y.J. & Prabhakar, N.R. (2003) Reactive oxygen species in the plasticity of respiratory behavior elicited by chronic intermittent hypoxia. *J. Appl. Physiol.* **94**, 2342–2349
33. Mitchell, G.S., Baker, T.L., Nanda, S.A., Fuller, D.D., Zabka, A.G., Hodgeman, B.A., Bavis, R.W., Mack, K.J., Olson, Jr, E.B. (2001) Intermittent hypoxia and respiratory plasticity. *J. Appl. Physiol.* **90**, 2466–2475.
34. Kumar, G.K., Rai, V., Sharma, S.D., Ramakrishnan, D.P., Peng, Y.J., Souvannakitti, D. & Prabhakar, N.R. (2006) Chronic intermittent hypoxia induces hypoxia-evoked catecholamine efflux in adult rat adrenal medulla via oxidative stress. *J. Physiol.* **575**, 229–239
35. Ziegler, M.G., Nelesen, R., Mills, P., Ancoli-Israel, S., Kennedy, B. & Dimsdale, J.E.(1997) Sleep apnea, norepinephrine-release rate, and daytime hypertension. *Sleep* **20**, 224–231
36. Ryan, S., Taylor, C.T. & McNicholas, W.T. (2006) Predictors of elevated nuclear factor-κB-dependent genes in obstructive sleep apnea syndrome. *Am. J. Respir. Crit. Care Med.* **174**, 824–830
37. Ryan, S., Taylor, C.T. & McNicholas, W.T. (2005) Selective activation of inflammatory pathways by intermittent hypoxia in obstructive sleep apnea syndrome. *Circulation* **112**, 2660–2667
38. Greenberg, H., Ye, X., Wilson, D., Htoo, A.K., Hendersen, T. & Liu, S.F. (2006) Chronic intermittent hypoxia activates nuclear factor-κB in cardiovascular tissues *in vivo*. *Biochem. Biophys. Res. Commun.* **343**, 591–596
39. Douglas, R.M., Xue, J., Chen, J.Y., Haddad, C.G., Alper, S.L. & Haddad, G.G. (2003) Chronic intermittent hypoxia decreases the expression of Na/H exchangers and HCO_3-dependent transporters in mouse CNS. *J. Appl. Physiol.* **95**, 292–299
40. Xu, W., Chi, L., Row, B.W., Xu, R., Ke, Y., Xu, B., Luo, C., Kheirandish, L., Gozal, D. & Liu, R. (2004) Increased oxidative stress is associated with chronic intermittent hypoxia-mediated brain cortical neuronal cell apoptosis in a mouse model of sleep apnea. *Neuroscience* **126**, 313–323
41. Gozal, E., Gozal, D., Pierce, W.M., Thongboonkerd, V., Scherzer, J.A., Sachleben, Jr, L.R., Brittian, K.R., Guo, S.Z., Cai, J. & Klein, J.B. (2002) Proteomic analysis of CA1 and CA3 regions of rat hippocampus and differential susceptibility to intermittent hypoxia. *J. Neurochem.* **83**, 331–345
42. Klein, J.B., Gozal, D., Pierce, W.M., Thongboonkerd, V., Scherzer, J.A., Sachleben, L.R., Guo, S.Z., Cai, J. & Gozal, E. (2003) Proteomic identification of a novel protein regulated in CA1 and CA3 hippocampal regions during intermittent hypoxia. *Respir. Physiol. Neurobiol.* **136**, 91–103
43. Li, R.C., Row, B.W., Gozal, E., Kheirandish, L., Fan, Q., Brittian, K.R., Guo, S.Z., Sachleben, Jr, L.R. & Gozal, D. (2003) Cyclooxygenase 2 and intermittent hypoxia-induced spatial deficits in the rat. *Am. J. Respir. Crit. Care Med.* **168**, 469–475
44. Goldbart, A., Row, B.W., Kheirandish, L., Schurr, A., Gozal, E., Guo, S.Z., Payne, R.S., Cheng, Z., Brittian, K.R. & Gozal, D. (2003) Intermittent hypoxic exposure during light phase induces changes

in cAMP response element binding protein activity in the rat CA1 hippocampal region: water maze performance correlates. *Neuroscience* **122**, 585–590

45. Gozal, E., Sachleben, Jr, L.R., Rane, M.J., Vega, C. & Gozal, D. (2005) Mild sustained and intermittent hypoxia induce apoptosis in PC-12 cells via different mechanisms. *Am. J. Physiol. Cell Physiol.* **288**, C535–C542
46. Kumar, G.K., Kim, D.K., Lee, M.S., Ramachandran, R. & Prabhakar, N.R. (2003) Activation of tyrosine hydroxylase by intermittent hypoxia: involvement of serine phosphorylation. *J. Appl. Physiol.* **95**, 536–544
47. Gozal, E., Shah, Z.A., Pequignot, J.M., Pequignot, J., Sachleben, Jr, L.R., Czyzyk-Krzeska, M.F., Li, R.C., Guo, S.Z. & Gozal, D. (2005) Tyrosine hydroxylase expression and activity in the rat brain: differential regulation after long-term intermittent or sustained hypoxia. *J. Appl. Physiol.* **99**, 642–649
48. Gardner, P.R., Nguyen, D.D. & White, C.W. (1994) Aconitase is a sensitive and critical target of oxygen poisoning in cultured mammalian cells and in rat lungs. *Proc. Natl. Acad. Sci. U.S.A.* **91**, 12248–12252
49. Guzy, R.D., Hoyos, B., Robin, E., Chen, H., Liu, L., Mansfield, K.D., Simon, M.C., Hammerling, U. & Schumacker, P.T. (2005) Mitochondrial complex III is required for hypoxia-induced ROS production and cellular oxygen sensing. *Cell Metab.* **1**, 401–408
50. Yuan, G., Adhikary, G. & Prabhakar, N.R. (2002) Activation of mitogen activated protein kinases by episodic hypoxia. Evidence for the involvement of reactive oxygen species. *FASEB J.* **16**, (Abstract)
51. Prabhakar, N.R. & Kline, D.D. (2002) Ventilatory changes during intermittent hypoxia: importance of pattern and duration. *High Alt. Med. Biol.* **3**, 195–204

8

Vascular adaptations to hypoxia: molecular and cellular mechanisms regulating vascular tone

Michael L. Paffett and Benjimen R. Walker[1]

Vascular Physiology Group, Department of Cell Biology and Physiology, University of New Mexico, MSC08 4750 Albuquerque, NM 87131, U.S.A.

Abstract

Several molecular and cellular adaptive mechanisms to hypoxia exist within the vasculature. Many of these processes involve oxygen sensing which is transduced into mediators of vasoconstriction in the pulmonary circulation and vasodilation in the systemic circulation. A variety of oxygen-responsive pathways, such as HIF (hypoxia-inducible factor)-1 and HOs (haem oxygenases), contribute to the overall adaptive process during hypoxia and are currently an area of intense research. Generation of ROS (reactive oxygen species) may also differentially regulate vascular tone in these circulations. Potential candidates underlying the divergent responses between the systemic and pulmonary circulations may include Nox (NADPH oxidase)-derived ROS and mitochondrial-derived ROS. In addition to alterations in ROS production governing vascular tone in the hypoxic setting, other vascular adaptations are likely to be involved. HPV (hypoxic pulmonary vasoconstriction) and CH (chronic hypoxia)-induced alterations in cellular proliferation, ionic conductances and changes in the contractile apparatus sensitivity to calcium, all occur as adaptive processes within the vasculature.

[1]*To whom correspondence should be addressed (email bwalker@salud.unm.edu).*

Introduction

The response to decreased PO_2 (partial pressure of oxygen) or hypoxia within the vasculature is complex and associated with many acute and chronic adaptations. In addition, the pulmonary and systemic circulations respond very differently to hypoxia although they may share common cellular signalling events initiated by low PO_2. Arteries from both circulations contain VSM (vascular smooth muscle) that dictates the active tone of the vessels and a metabolically active endothelium. Due to its proximity to the endothelium, the VSM receives paracrine cues in response to decreasing PO_2, as well as sensing hypoxia directly through various cellular mechanisms. The endothelium not only serves as a protective barrier from the circulating blood, but also is a rich source of vasoactive substances that determine vascular tone under a variety of conditions, including hypoxia. A variety of molecular pathways and cellular sensors respond to hypoxia which activate homoeostatic mechanisms in the cardiovascular and respiratory systems. These oxygen-sensing capabilities within the vasculature often bring about change in gene expression and signal transduction pathways that uniquely determine the haemodynamic state within a given vascular bed that can be deemed adaptive in some circumstances and/or pathological in others depending upon the magnitude and duration of the hypoxic stimulus.

Vascular oxygen-sensing mechanisms

Hypoxia-inducible gene expression

One of the most studied transcription factors responsive to reduced PO_2 is HIF (hypoxia-inducible factor)-1. Analysis of purified HIF-1 revealed a heterodimeric protein containing HIF-1α and HIF-1β subunits also referred to as ARNT (aryl hydrocarbon receptor nuclear translocator). Consequently, protein structure analysis has demonstrated that each protein in this heterodimer contains a bHLH (basic helix-loop-helix) domain capable of binding to the promoter region and activating transcription of a variety of genes under hypoxic conditions [1]. However, under normoxic conditions the VHL (von Hippel-Lindau) tumour suppressor protein binds to two hydroxy-proline residues at Pro^{402} and Pro^{564} located in the transactivation domain of HIF-1α and this complex is then recognized by the E3 ubiquitin ligase which routes this complex through the 26S proteasomal degradation pathway (left-hand panel of Figure 1). Furthermore, a third hydroxylation site at Asp^{803} serves to inhibit CBP [CREB (cAMP-response-element-binding protein)-binding protein] and p300 binding, therefore restricting the activity of these transcriptional co-activators [2]. Hydroxylation of these residues is carried out by prolyl and aparaginyl hydroxylases in which molecular oxygen serves as the rate-limiting substrate for this reaction. Under hypoxic conditions hydroxylase activity subsides and the loss of hydroxylation at these residues results in a stable HIF-1, which is then capable of recruiting p300/CBP, binding to the promoter region

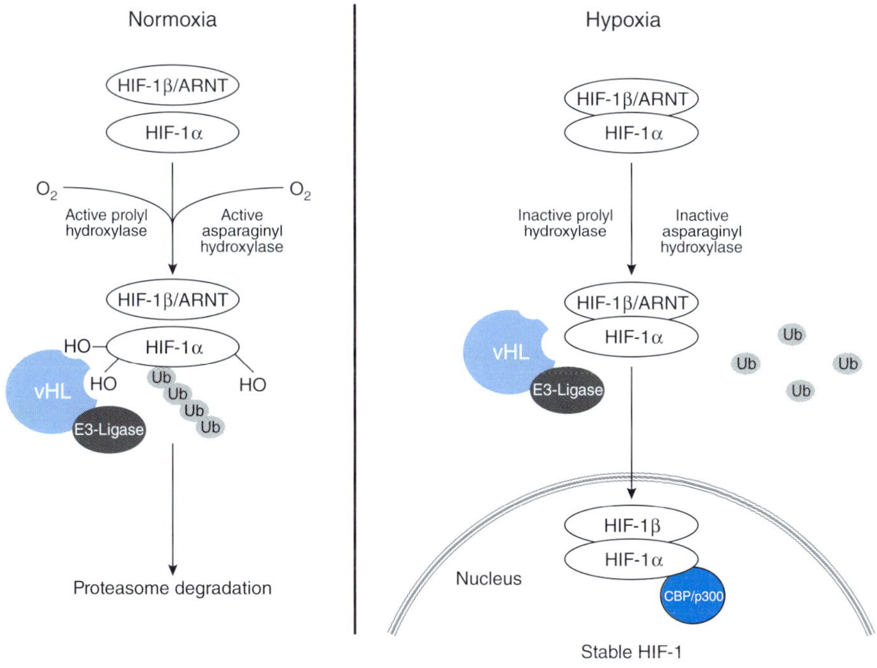

Figure 1. Mechanism of oxygen-dependent degradation of HIF-1
Oxygen-dependent prolyl hydroxylation allows the VHL and E3 ligase complex to associate with HIF-1 in order for the E3 ligase to ubiquitinate the HIF-1α subunit. Asparaginyl hydroxylation of HIF-1 prevents binding of the transcription factor p300/CBP. During hypoxic conditions, prolyl and asparaginyl hydroxylase activity is reduced, thus facilitating the translocation of this complex to the nucleus where HIF-1 then serves as a transcription factor. OH, hydroxylation sites for oxygen-dependent prolyl and asparaginyl hydroxylases; Ub, ubiquitin.

containing a HRE (hypoxia-response element) and inducing gene expression (right-hand panel of Figure 1). For example, numerous studies have shown HIF-1 dependent expression of ET-1 (endothelin-1) in the vascular endothelium resulting in elevated plasma ET-1 during hypoxic conditions [3]. In turn, ET-1 has pleiotropic effects in which it can cause vasoconstriction, alterations in VSM calcium sensitivity and proliferation of pulmonary VSM leading to increased vascular resistance. These alterations in vascular function associated with CH (chronic hypoxia) are discussed in more detail below. Table 1 lists a number of HIF-responsive genes and the proposed roles for the encoded protein within the vasculature. Furthermore, regulation of HIF-1α by agonist-induced ROS (reactive oxygen species) production may occur in both pulmonary and aortic VSM cells under non-hypoxic conditions, since pre-treatment with antioxidants averts HIF-1α accumulation and expression of HIF-1-dependent genes [4,5]. These findings clearly suggest a role for ROS in regulating HIF-1α activity and implicate ROS as a modulator of HIF-1 pathways within pulmonary and systemic arteries. Although these studies did not investigate the effects of hypoxia on ROS dependent HIF-1 stabilization, these observations do provide insight into HIF-1α regulation by ROS production.

Table 1. Partial list of HIF-1α-responsive genes and associated roles within the vasculature

Protein	Role(s)
Adrenergic receptor (β_1)	Apoptosis and cardiovascular development
ET-1	Vasoconstriction
	Vascular remodelling
	Altered VSM calcium sensitivity
HO-1	Vasodilation and ROS scavenging
Inducible NO synthase	Vasodilation and inflammation
Vascular endothelial growth factor	Angiogenesis and endothelial permeability

Interestingly, ROS production is also altered in the hypoxic setting [6]. The effects of hypoxia on ROS production and the accompanying divergent vasomotor responses between vascular beds will be discussed in the next section.

Nox (NADPH oxidase)

A rapidly emerging body of evidence suggests that Nox may be a putative oxygen sensor within the vasculature. Noxs consist of a membrane-bound phox (phagocytic oxidase), cytosolic regulatory subunits and a small GTPase. The flavin- and haem-binding glycoprotein (gp91*phox* or Nox2) and p22*phox* subunits constitute the membrane-associated flavocytochrome b_{558} initially discovered by spectral analysis. Binding of the regulatory cytosolic p40*phox*, p47*phox* and p67*phox* proteins increase the activity of Nox by enhancing flux through the core oxidase (Figure 2). The small GTP-binding protein Rac1/2 also imposes a regulatory function on Nox activity. Several Nox isoforms have been identified, of which Nox1–4 lack intracellular calcium-binding sites; however Nox5

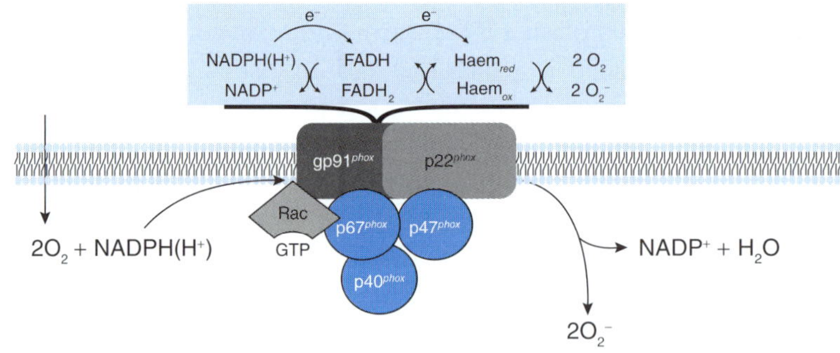

Figure 2. Nox and ancillary proteins that modulate electron flux generating superoxide anion

The core oxidase composition is composed of gp91*phox* and p22*phox* subunits located at the plasma membrane. GTP-bound Rac, p47*phox*, p40*phox* and p67*phox* have been identified as modulators of Nox activity. NADPH(H^+) and molecular oxygen serve as substrates for Nox-generating superoxide anion and $NADP^+$ and H_2O. The two-step electron transfer generating 2 moles of superoxide anion from NADPH(H^+) and FAD to convert $haem_{ox}$ to $haem_{red}$ in the gp91*phox* subunit is indicated in the inset.

has four calcium-binding EF hands near the N-terminus. The EF-hand motifs impart a calcium sensitivity to the enzyme and increases in cytosolic calcium augment ROS production [7]. Nox homologues of the first identified gp91^{phox} subunit possess a wide range of sequence similarities and are found in a variety of cell types between species and cultured cell lines. Additionally, the novel isoform of p47^{phox} termed Noxo1 (Nox organizer 1) has unique structural characteristics which include a deletion of an autoinhibitory domain and differs in its lipid-binding specificity compared with p47^{phox} [8]. Additionally, the p67^{phox} homologue, p51^{phox} or Noxa1 (NADPH oxidase activator 1) has a high degree of sequence homology to its parent isoform and regulates Nox activity by interacting with the flavocytochrome complex [9]. The existence of these two ancillary protein variants gives rise to uniquely composed Noxs which may offer further insight regarding Nox function. Interestingly, gp91^{phox} and the cytosolic regulators (p40^{phox,} p47^{phox} and p67^{phox}) exist in the VSM [10,11] and participate in ROS production following receptor stimulation. Nox requires molecular oxygen and NADPH as substrates which results in the generation of the superoxide anion (O_2^-) which is then capable of permeating the plasma membrane via anionic channels and being further reduced to H_2O_2 by SOD (superoxide dismutase). Either of these ROS can modify a variety of signal transduction pathways that regulate vascular tone and expression of specific genes known to affect vascular function. Hypoxic transitions were initially predicted to limit oxidant (i.e. O_2^- and H_2O_2) production and promote a more reduced state within the cytosol. Although this appears to be the case in systemic arteries which generally dilate in response to hypoxia, pulmonary arteries constrict and Nox activity appears to increase in this setting.

HO (haemoxygenase)

HOs represent another postulated cellular oxygen sensor within many different cell types, including the vasculature. HOs catalyse the oxidation of haem to equimolar parts of carbon monoxide, B*v* (biliverdin) and free ferrous iron (Figure 3). Three known isoforms, HO-1, HO-2 and HO-3 have been identified to date. HO-1 is largely considered to be the only inducible isoform. HO-1 expression is associated with elevated oxidative stress, shear stress, as well as with hypoxia. HO-2 is constitutively expressed in neurons and the vascular endothelium, with only a limited number of physiological activators, including adrenal glucocorticoids and protein kinase C. The most recently identified isoform, HO-3, appears to be incapable of haem degradation and is thought to primarily serve as a haem-sensing and/or -binding protein [12]. Since HOs require oxygen to degrade haem, they are theorized to be putative oxygen sensors. The majority of attention within the field of vascular physiology centres on the vasodilatory role of the HO product, carbon monoxide, on vascular tone, specifically within the VSM. Carbon monoxide acts upon the haem-containing protein sGC (soluble guanylate cyclase) to stimulate cyclic GMP production [13] leading to protein kinase G-dependent vasodilation in systemic arterial beds [14]. Furthermore, the

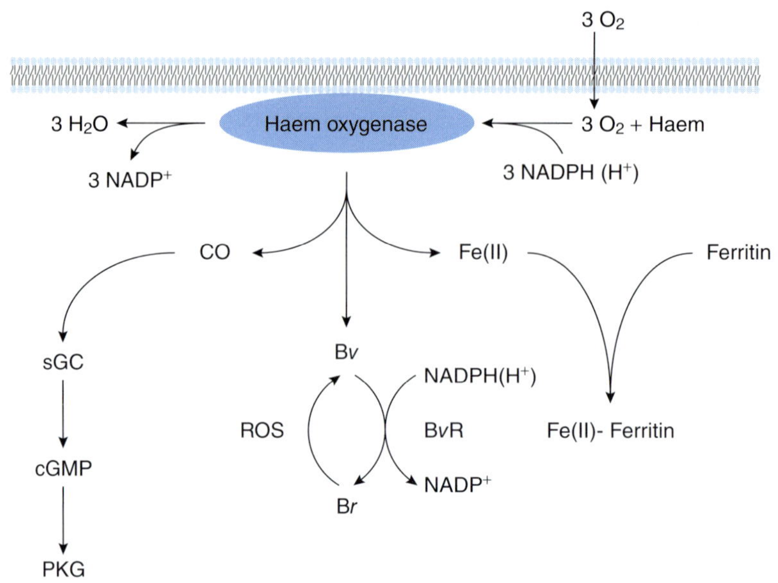

Figure 3. Role of HO in oxygen-sensing and vascular control
HOs utilize haem and molecular oxygen to generate carbon monoxide, B*r* and Fe(II). Carbon monoxide can then activate sGC to elicit protein kinase G-dependent relaxation. B*v* is reduced to the potent anti-oxidant B*r* by B*v*R. Fe(II) sequestration occurs by binding to ferritin.

potent vasodilator NO (nitric oxide), produced by the vascular endothelium also causes vasodilation by stimulating sGC activity in a similar fashion. Interestingly, increases in HO-1 expression occur following Nox inhibition, presumably by signalling via protein kinase B and p38 MAPK (mitogen-activated protein kinase) signalling pathways [15]. These results suggest that HO-1 expression is sensitive to a decrease in Nox-associated ROS formation and are consistent with similar findings where HO-1 expression increased after treatment with the anti-oxidant pyrrolidine dithiocarbamate [16]. Although these findings suggest up-regulation of HO-1 gene expression in monocytes by scavenging endogenous ROS, rat aortic VSM cells show a ROS-dependent increase in HO-1 expression through activation of p42/44 MAPK pathways and AP-1 (activator protein-1) [17]. These discrepancies in HO-1 gene expression may be due to tissue-specific differences in redox-sensitive transcriptional pathways or non-specific effects of the ROS-scavenging compounds utilized in previous studies. In addition to the importance of redox state and HO-1 transcriptional activity, the conversion of B*v* to the potent anti-oxidant B*r* (bilirubin) by B*v*R (B*v* reductase) is hypothesized to be an efficient ROS scavenging system [18]. The accumulation of Br provides an anti-oxidant pool to scavenge endogenous ROS. Several studies demonstrate the protective nature of HO-1 induction via pharmacological or gene transfer techniques in which blood pressure restoration and endothelial oxidative injury were curtailed in renal models of hypertension and hyperglycaemic rats [19]. An interesting aspect of haem degradation is that under conditions that induce HO-1, such as hypoxia, this putative oxygen sensor may provide protective antioxidant effects against increased ROS formation associated with various acute and chronic disease states.

Mitochondria as oxygen sensors

Cellular respiration requires that molecular oxygen serves as a terminal electron sink in the distal electron transport chain to establish a proton-motive force necessary for ATP production. Similar to Nox, mitochondria generate ROS as metabolic by-products. As NADH and FADH are generated from the tricarboxylic acid cycle, electrons are transferred through iron-sulfur centres located in the electron transport chain. During this process of oxidation and reduction, O_2^- is produced at several points along the electron transport chain. ROS production occurs at complex I and at two distinct redox sites of the quinone cycle at complex III. Although measurement of basal mitochondrial ROS production is difficult, from mitochondria in pulmonary arteries conflicting evidence exists regarding the effect of hypoxia on this variable. Various levels of mitochondrial SODs have also been investigated in the systemic and pulmonary arterial circulations and are proposed to influence the level of ROS production in these vascular beds [20]. It has also been postulated that during hypoxia a build-up of NADH ensues due to the reduction in the terminal electron acceptor, molecular oxygen [21]. These clues may be useful in determining the overall mechanisms responsible for the different responses to hypoxia in these vascular beds.

Hypoxia: ROS and redox state

The role of ROS during hypoxia

Although it has been a long-held belief that ROS are detrimental to cellular physiology, more recent evidence suggests that ROS are critical mediators in many signalling pathways. As an adaptive process to hypoxia, reduced PO_2 must be sensed and then transduced by an effector molecule that brings about change within a system. It is typically thought that systemic arteries dilate and pulmonary arteries constrict in response to hypoxia. Given the likelihood of similar oxygen-sensing capabilities between tissues, unique downstream effectors may underlie these divergent responses. This may be due to fundamental differences in ROS production between these circulations under normoxic conditions (Table 2). Interestingly, basal production of NADPH is greater in bovine pulmonary arteries than in coronary arteries presumably due to greater expression of G6PD (glucose-6-phosphate dehydrogenase) [22]. In contrast to the pulmonary circulation, it is hypothesized that normoxic NADPH production and G6PD expression are diminished in systemic arteries by competition between glycolytic and pentose phosphate pathways [23]. Concomitant decreases in O_2^- and H_2O_2

Table 2. Divergent effects of hypoxia on ROS production in the pulmonary and systemic circulations

Arterial bed	ROS source	Vasomotor effect
	Mitochondrial	Extra-mitochondrial
Pulmonary	Increase/decrease	Increase Vasoconstriction
Systemic	Increase	Decrease Vasodilation

production promoting relaxation are also observed in this systemic vascular bed [20]. Furthermore, these investigators hypothesize that the decrease in basal NADPH production may in turn be responsible for the accompanying vasodilation observed in systemic vascular beds during hypoxia. This diminished availability of substrate (NADPH) for Nox may be a critical variable that distinguishes coronary artery vasodilation from pulmonary artery vasoconstriction during hypoxia. The differences in basal ROS production between these circulations may provide insight into the contrasting hypoxic vasomotor responses commonly observed.

A general assumption in redox chemistry is that when PO_2 decreases, a subsequent reduction in free-radical generation follows by virtue of the fact that there is less substrate to generate ROS. However, this prediction is not observed in pulmonary VSM cells in which increased NADPH-dependent superoxide formation is observed [24] during hypoxia. Furthermore, recent evidence suggests that CH may actually lead to increased O_2^- production, presumably by increased activity of NADPH oxidase [25]. These investigators observed enhanced agonist-induced vasoconstriction and increased muscularization of intrapulmonary arteries following CH in gp91^{phox} wild-type mice compared with the gp91^{phox} knockout independent of any changes in Nox-subunit mRNA expression. The divergent vasomotor responses to hypoxia between the pulmonary and coronary (systemic) circulations appear to be associated with fundamentally unique metabolic and signalling pathways that account for differential regulation of vascular resistance in response to hypoxia between these beds.

In addition to various responses of Nox during hypoxia, conflicting effects of hypoxia on mitochondrial ROS generation in pulmonary arteries may exist. For example, hypoxia has been shown to augment ROS production and cause vasoconstriction that is attenuated by the distal complex III inhibitor myxothiozol [26]. In addition to redox inhibition of the mitochondrial electron transport chain, others have demonstrated the loss of contractile response to hypoxia following the depletion of essential electron transport chain subunits with a two week exposure to ethidium bromide [27]. It has been further hypothesized that a decrease in hypoxia-induced ROS production leads to a change in the redox equilibrium toward a more reduced state. In contrast, others have observed a decreased ROS production during hypoxia or following complex I and proximal complex III inhibition [28,29]. Therefore conflicting data exist regarding the influence of hypoxia on ROS production and the role of these mitochondrial and extra-mitochondrial effector pathways in the regulation of vascular tone.

Effect of redox state on VSM potassium conductances

As discussed above, hypoxia appears to differentially impact the production of ROS and perturb the equilibrium of cellular reducing equivalents such as NADPH. The primary agent for maintaining redox homoeostasis is glutathione. This compound can exist either in the reduced (GSH) or oxidized state (GSSG). The reduced form of glutathione with glutathione peroxidase are critical components that maintain appropriate levels of endogenous peroxides. As part of the regenerative cycle to replenish GSH from GSSG, glutathione reductase activity is largely determined from cytosolic NADPH concentrations. The associated changes in cellular metabolism associated with the cytosolic NADPH(H^+)/$NADP^+$ ratio and ROS production during hypoxia are important variables that can alter redox homoeostasis and ultimately lead to functional changes within the vasculature. For example, the Kv (voltage-sensitive potassium) channels Kv1.5 and Kv2.1 have been implicated in HPV (hypoxic pulmonary vasoconstriction) [30]. It is hypothesized that hypoxia increases the reduced state within the VSM resulting in a decreased open probability of these channels leading to depolarization of the sarcolemma and activation of L type VGCCs (voltage-gated calcium channels). The increase in open probability of VGCCs results in an increased global VSM calcium concentration and constriction of the artery. Although this potential effect of hypoxia on ROS production and an overall increase in the cellular reducing state in pulmonary arteries results in vasoconstriction, renal arteries show increased ROS production and exhibit vasodilation suggesting increased outward potassium conductances in this setting [20]. Furthermore, potassium channel conductances in the ductus arteriosis are thought to decrease in the face of increasing PO_2 [31]. These examples further demonstrate the differential effects of hypoxia and reducing state in pulmonary and systemic arterial bed. The proposed role of altered cellular redox state to differentially influence potassium channels and thus vasomotor responses between the pulmonary and systemic arterial circulations is intriguing, however this picture is still not complete.

Effects of acute hypoxia in pulmonary arteries

HPV

The pulmonary circulation is typically characterized by its low resistance, low pressure profile that is advantageous for gas exchange between the alveoli and surrounding capillary beds. Even under normal conditions there is a small proportion of respiratory airways that are not adequately ventilated and therefore local regions of hypoxia develop. The response to this localized decrease in PO_2 is pulmonary arterial vasoconstriction that diverts blood flow to better ventilated areas, thereby matching ventilation and perfusion. However, global hypoxia observed in many pulmonary disease states and presence at high altitude leads to generalized pulmonary artery vasoconstriction and the eventual development of pulmonary hypertension. This response is unique to the pulmonary arterial circulation and is conserved in the respiratory organs of a variety of phylogenetically unrelated species, demonstrating its physiological importance in regulating gas exchange. Many hypotheses and potential mechanisms have been proposed to explain this phenomenon such as redox inhibition of Kv channels leading to VGCC-dependent calcium influx and contraction (described above). However, other potential mechanisms for HPV may also contribute to the response. For example, hypoxia-induced calcium release from the sarcoplasmic reticulum of VSM cells has been linked to increased cyclic ADP ribose levels in small pulmonary arteries [32]. It is proposed that cyclic ADP ribose stimulates the release of calcium by ryanodine receptors. Cyclic ADP ribose stability appears to be regulated by cytosolic NADH levels and levels of this reduced compound are elevated during acute hypoxia. Others have shown a link between lowering of VSM intracellular calcium stores and subsequent activation of store-operated calcium entry in intrapulmonary arteries [33]. Furthermore, removal of the endothelium reduces HPV by inhibition of Rho kinase sensitization of the VSM contractile apparatus to calcium [34]. These results suggest that hypoxia elicits the release of an endothelium-derived substance that modulates the sensitivity of the contractile apparatus to calcium in the underlying VSM. ET-1 may be the endothelium-derived substance responsible for this effect since administration of exogenous ET-1 restores HPV following endothelial disruption [35]. It is clear from the above discussion and numerous other reports in the literature, that HPV may result from the interaction of several signalling pathways.

Effects of CH in pulmonary arteries

Conditions associated with CH, such as COPD (chronic obstructive pulmonary disease) or prolonged residence at high altitude, result in several vascular adaptations within the arterial segment of the lung that contribute to the development of pulmonary hypertension. These changes include vascular remodelling of the arterial segment as well as enhanced vasoconstrictor reactivity, both of which increase pulmonary vascular resistance leading to hypertension.

Structural alterations: remodelling of the VSM

Medial and intimal remodelling in response to CH increases pulmonary vascular resistance by reducing the luminal diameter of the vessel. This encroachment is characterized by increased growth and proliferation of the VSM. As previously discussed, a key factor in CH-induced arterial VSM remodelling appears to centre on HIF-dependent transcriptional up-regulation of ET-1. ET-1 has many effects on the underlying VSM. For example, upon binding to VSM ET_A receptors, ET-1 causes a rise in VSM intracellular calcium and subsequent vasoconstriction. In addition to evoking acute vasoconstriction, elevated VSM calcium also activates the calcium-sensitive transcription factor CBP resulting in up-regulation of growth specific genes [36]. This in turn leads to increased wall thickening and muscularization of previously thin-walled arteries in the lung [37]. These proliferative responses are also seen in the intimal layer (endothelium) in patients with mild COPD [38]. This proliferative response of the endothelium has been correlated to a higher degree of endothelial dysfunction [39], thereby leading to a more hypertensive state within the lungs of COPD patients. In addition to the hypertrophic response of the VSM and endothelium, increased deposition of the ECM (extracellular matrix) proteins collagen and elastin may result through greater expression of TGF-β (transforming growth factor-β). The role of TGF-β has been examined where a dominant-negative mutation in the TGF-β receptor impairs the development of CH-induced pulmonary hypertension, presumably by limiting arterial remodelling [40]. The phenotypic composition of the surrounding adventitia is also altered [41] in which the proportion of fibrobast sub-populations shifts to a myofibroblast dominating phenotype. Thus it is apparent that hyperplasia and hypertrophy of the cellular elements of the vascular wall leads to structural alterations that influence the degree of pulmonary hypertension in the setting of CH and/or COPD.

VSM depolarization and decreased Kv channel expression

E_m (resting membrane potential) is an efficient regulator of calcium influx as discussed above in the case of the redox effects on Kv channel activity during HPV. Interestingly, VSM from small pulmonary arteries exposed to CH as well as VSM exposed to CH in culture have a more depolarized E_m than controls [42,43], thus leading to increased calcium influx and vasoconstriction. It appears that this depolarization is due to down-regulation of several isoforms of Kv channels in the VSM [44]. Furthermore, inhibition of potassium channel activity and/or expression may also inhibit apoptosis [45], thus potentially contributing to the more growth-responsive state observed in remodelling of the VSM during CH. Therefore decreased potassium channel expression during CH appears to alter the VSM E_m but also enhances vascular remodelling in this setting.

Alterations in calcium sensitivity of the contractile apparatus

The small GTPase RhoA and its downstream effector Rho kinase have gained considerable attention with regard to the pathogenesis of CH-induced pulmonary hypertension. The RhoA/Rho kinase pathway plays an important role in regulating the contractile state of VSM by affecting the degree of MLC (myosin light chain) phosphorylation through inhibition of MLC phosphatase activity. Regulation of cross-bridge cycling is governed by a balance between activities of the calcium/calmodulin-dependent MLC kinase and calcium-independent MLC phosphatase. Activation of the RhoA/Rho kinase pathway through ligand-dependent and ligand-independent mechanisms plays a significant role in the sustained vasoconstriction following CH. For example, Rho kinase inhibition, but not L-type VGCC inhibition elicits vasodilation in lungs from CH rats, suggesting increased VSM calcium sensitivity contributes to pulmonary hypertension in this setting [46,47].

Conclusion

Although the oxygen-sensing mechanisms outlined above exist in both the pulmonary and systemic vasculatures, the specific means by which they differentially affect vascular tone is still unclear. The role of these putative mechanisms in hypoxia-induced systemic vasodilation and pulmonary vasoconstriction continues to be an area of intense research and debate. Advances in our understanding of vascular adaptations to hypoxia will most likely centre on the role of redox state in oxygen-sensing within the vasculature, and how ROS influences signalling pathways that determine vascular tone.

Summary

- *A variety of oxygen-sensing mechanisms exist that alter the reactivity of the vasculature.*
- *ROS production from various sites within the mitochondrial electron transport chain as well as extra-mitochondrial sources appear to play unique roles in pulmonary and systemic arteries.*
- *The redox state of the vessel wall may vary between these two circulations and these differences may provide further insight into how each circulation responds and adapts to hypoxia.*
- *A variety of structural and electrophysiological alterations take place following CH in the pulmonary vasculature that contributes to the development of pulmonary hypertension.*

References

1. Wenger, R.H. (2002) Cellular adaptation to hypoxia: O_2-sensing protein hydroxylases, hypoxia-inducible transcription factors, and O_2-regulated gene expression. *FASEB J.* **16**, 1151–1162
2. Lando, D., Peet, D.J., Gorman, J.J., Whelan, D.A., Whitelaw, M.L. & Bruick, R.K. (2002) FIH-1 is an asparaginyl hydroxylase enzyme that regulates the transcriptional activity of hypoxia-inducible factor. *Genes Dev.* **16**, 1466–1471
3. Semenza, G.L. (2000) Oxygen-regulated transcription factors and their role in pulmonary disease. *Respir. Res.* **1**, 159–162
4. BelAiba, R.S., Djordjevic, T., Bonello, S., Flugel, D., Hess, J., Kietzmann, T. & Gorlach, A. (2004) Redox-sensitive regulation of the HIF pathway under non-hypoxic conditions in pulmonary artery smooth muscle cells. *Biol. Chem.* **385**, 249–257
5. Gorlach, A., Diebold, I., Schini-Kerth, V.B., Berchner-Pfannschmidt, U., Roth, U., Brandes, R.P., Kietzmann, T. & Busse, R. (2001) Thrombin activates the hypoxia-inducible factor-1 signaling

pathway in vascular smooth muscle cells: role of the p22(phox)-containing NADPH oxidase. *Circ. Res.* **89**, 47–54

6. Wolin, M.S., Ahmad, M. & Gupte, S.A. (2005) Oxidant and redox signaling in vascular oxygen sensing mechanisms: basic concepts, current controversies, and potential importance of cytosolic NADPH. *Am. J. Physiol. Lung Cell. Mol. Physiol.* **289**, L159–L173
7. Banfi, B., Tirone, F., Durussel, I., Knisz, J., Moskwa, P., Molnar, G.Z., Krause, K.H. & Cox, J.A. (2004) Mechanism of Ca^{2+} activation of the NADPH oxidase 5 (NOX5). *J. Biol. Chem.* **279**, 18583–18591
8. Cheng, G., Ritsick, D. & Lambeth, J.D. (2004) Nox3 regulation by NOXO1, p47phox, and p67phox. *J. Biol. Chem.* **279**, 34250–34255
9. Geiszt, M., Lekstrom, K., Witta, J. & Leto, T.L. (2003) Proteins homologous to p47phox and p67phox support superoxide production by NAD(P)H oxidase 1 in colon epithelial cells. *J. Biol. Chem.* **278**, 20006–20012
10. Lassegue, B., Sorescu, D., Szocs, K., Yin, Q., Akers, M., Zhang, Y., Grant, S.L., Lambeth, J.D. & Griendling, K.K. (2001) Novel gp91(phox) homologues in vascular smooth muscle cells: nox1 mediates angiotensin II-induced superoxide formation and redox-sensitive signaling pathways. *Circ. Res.* **88**, 888–894
11. Patterson, C., Ruef, J., Madamanchi, N.R., Barry-Lane, P., Hu, Z., Horaist, C., Ballinger, C.A., Brasier, A.R., Bode, C. & Runge, M.S. (1999) Stimulation of a vascular smooth muscle cell NAD(P)H oxidase by thrombin. Evidence that p47(phox) may participate in forming this oxidase *in vitro* and *in vivo*. *J. Biol. Chem.* **274**, 19814–19822
12. McCoubrey, Jr, W.K., Huang, T.J. & Maines, M.D. (1997) Isolation and characterization of a cDNA from the rat brain that encodes hemoprotein heme oxygenase-3. *Eur. J. Biochem.* **247**, 725–732
13. Morita, T., Perrella, M.A., Lee, M.E. & Kourembanas, S. (1995) Smooth muscle cell-derived carbon monoxide is a regulator of vascular cGMP. *Proc. Natl. Acad. Sci. U.S.A.* **92**, 1475–1479
14. Gagov, H., Kadinov, B., Hristov, K., Boev, K., Itzev, D., Bolton, T. & Duridanova, D. (2003) Role of constitutively expressed heme oxygenase-2 in the regulation of guinea pig coronary artery tone. *Pflugers Arch.* **446**, 412–421
15. Wijayanti, N., Kietzmann, T. & Immenschuh, S. (2005) Heme oxygenase-1 gene activation by the NAD(P)H oxidase inhibitor 4-(2-aminoethyl) benzenesulfonyl fluoride via a protein kinase B, p38-dependent signaling pathway in monocytes. *J. Biol. Chem.* **280**, 21820–21829
16. Hartsfield, C.L., Alam, J. & Choi, A.M. (1998) Transcriptional regulation of the heme oxygenase 1 gene by pyrrolidine dithiocarbamate. *FASEB J.* **12**, 1675–1682
17. Cheng, P.Y., Lee, Y.M., Shih, N.L., Chen, Y.C. & Yen, M.H. (2006) Heme oxygenase-1 contributes to the cytoprotection of α-lipoic acid via activation of p44/42 mitogen-activated protein kinase in vascular smooth muscle cells. *Free Radical Biol. Med.* **40**, 1313–1322
18. Matsumoto, H., Ishikawa, K., Itabe, H. & Maruyama, Y. (2006) Carbon monoxide and bilirubin from heme oxygenase-1 suppresses reactive oxygen species generation and plasminogen activator inhibitor-1 induction. *Mol. Cell Biochem.* **291**, 21–28
19. Abraham, N.G. & Kappas, A. (2005) Heme oxygenase and the cardiovascular-renal system. *Free Radical Biol. Med.* **39**, 1–25
20. Michelakis, E.D., Hampl, V., Nsair, A., Wu, X., Harry, G., Haromy, A., Gurtu, R. & Archer, S.L. (2002) Diversity in mitochondrial function explains differences in vascular oxygen sensing. *Circ. Res.* **90**, 1307–1315
21. Freeman, B.A. & Crapo, J.D. (1981) Hyperoxia increases oxygen radical production in rat lungs and lung mitochondria. *J. Biol. Chem.* **256**, 10986–10992
22. Gupte, S.A., Kaminski, P.M., Floyd, B., Agarwal, R., Ali, N., Ahmad, M., Edwards, J. & Wolin, M.S. (2005) Cytosolic NADPH may regulate differences in basal Nox oxidase-derived superoxide generation in bovine coronary and pulmonary arteries. *Am. J. Physiol Heart Circ. Physiol.* **288**, H13–H21
23. Gupte, S.A., Arshad, M., Viola, S., Kaminski, P.M., Ungvari, Z., Rabbani, G., Koller, A. & Wolin, M.S. (2003) Pentose phosphate pathway coordinates multiple redox-controlled relaxing mechanisms in bovine coronary arteries. *Am. J. Physiol Heart Circ. Physiol.* **285**, H2316–H2326

24. Marshall, C., Mamary, A.J., Verhoeven, A.J. & Marshall, B.E. (1996) Pulmonary artery NADPH-oxidase is activated in hypoxic pulmonary vasoconstriction. *Am. J. Respir. Cell Mol. Biol.* **15**, 633–644
25. Liu, J.Q., Zelko, I.N., Erbynn, E.M., Sham, J.S. & Folz, R.J. (2006) Hypoxic pulmonary hypertension: role of superoxide and NADPH oxidase (gp91phox). *Am. J. Physiol. Lung Cell Mol. Physiol.* **290**, L2–L10
26. Waypa, G.B., Guzy, R., Mungai, P.T., Mack, M.M., Marks, J.D., Roe, M.W. & Schumacker, P.T. (2006) Increases in mitochondrial reactive oxygen species trigger hypoxia-induced calcium responses in pulmonary artery smooth muscle cells. *Circ. Res.* **99**, 970–978
27. Waypa, G.B., Chandel, N.S. & Schumacker, P.T. (2001) Model for hypoxic pulmonary vasoconstriction involving mitochondrial oxygen sensing. *Circ. Res.* **88**, 1259–1266
28. Archer, S.L., Huang, J., Henry, T., Peterson, D. & Weir, E.K. (1993) A redox-based O_2 sensor in rat pulmonary vasculature. *Circ. Res.* **73**, 1100–1112
29. Paky, A., Michael, J.R., Burke-Wolin, T.M., Wolin, M.S. & Gurtner, G.H. (1993) Endogenous production of superoxide by rabbit lungs: effects of hypoxia or metabolic inhibitors. *J. Appl. Physiol.* **74**, 2868–2874
30. Archer, S.L., Wu, X.C., Thebaud, B., Nsair, A., Bonnet, S., Tyrrell, B., McMurtry, M.S., Hashimoto, K., Harry, G. & Michelakis, E.D. (2004) Preferential expression and function of voltage-gated, O_2-sensitive K^+ channels in resistance pulmonary arteries explains regional heterogeneity in hypoxic pulmonary vasoconstriction: ionic diversity in smooth muscle cells. *Circ. Res.* **95**, 308–318
31. Olschewski, A., Hong, Z., Peterson, D.A., Nelson, D.P., Porter, V.A. & Weir, E.K. (2004) Opposite effects of redox status on membrane potential, cytosolic calcium, and tone in pulmonary arteries and ductus arteriosus. *Am. J. Physiol. Lung Cell Mol. Physiol.* **286**, L15–L22
32. Wilson, H.L., Dipp, M., Thomas, J.M., Lad, C., Galione, A. & Evans, A.M. (2001) ADP-ribosyl cyclase and cyclic ADP-ribose hydrolase act as a redox sensor: a primary role for cyclic ADP-ribose in hypoxic pulmonary vasoconstriction. *J. Biol. Chem.* **276**, 11180–11188
33. Snetkov, V.A., Aaronson, P.I., Ward, J.P., Knock, G.A. & Robertson, T.P. (2003) Capacitative calcium entry as a pulmonary specific vasoconstrictor mechanism in small muscular arteries of the rat. *Br. J. Pharmacol.* **140**, 97–106
34. Robertson, T.P., Aaronson, P.I. & Ward, J.P. (2003) Ca^{2+} sensitization during sustained hypoxic pulmonary vasoconstriction is endothelium dependent. *Am. J. Physiol. Lung Cell Mol. Physiol.* **284**, L1121–L1126
35. Liu, Q., Sham, J.S., Shimoda, L.A. & Sylvester, J.T. (2001) Hypoxic constriction of porcine distal pulmonary arteries: endothelium and endothelin dependence. *Am. J. Physiol. Lung Cell Mol. Physiol.* **280**, L856–L865
36. Stenmark, K.R. & Mecham, R.P. (1997) Cellular and molecular mechanisms of pulmonary vascular remodeling. *Annu. Rev. Physiol.* **59**, 89–144
37. Klemm, D.J., Watson, P.A., Frid, M.G., Dempsey, E.C., Schaack, J., Colton, L.A., Nesterova, A., Stenmark, K.R. & Reusch, J.E. (2001) cAMP response element-binding protein content is a molecular determinant of smooth muscle cell proliferation and migration. *J. Biol. Chem.* **276**, 46132–46141
38. Barbera, J.A., Riverola, A., Roca, J., Ramirez, J., Wagner, P.D., Ros, D., Wiggs, B.R. & Rodriguez-Roisin, R. (1994) Pulmonary vascular abnormalities and ventilation-perfusion relationships in mild chronic obstructive pulmonary disease. *Am. J. Respir. Crit. Care Med.* **149**, 423–429
39. Peinado, V.I., Barbera, J.A., Ramirez, J., Gomez, F.P., Roca, J., Jover, L., Gimferrer, J.M. & Rodriguez-Roisin, R. (1998) Endothelial dysfunction in pulmonary arteries of patients with mild COPD. *Am. J. Physiol.* **274**, L908–L913
40. Chen, Y.F., Feng, J.A., Li, P., Xing, D., Zhang, Y., Serra, R., Ambalavanan, N., Majid-Hassan, E. & Oparil, S. (2006) Dominant negative mutation of the TGF-β receptor blocks hypoxia-induced pulmonary vascular remodeling. *J. Appl. Physiol.* **100**, 564–571
41. Stenmark, K.R., Gerasimovskaya, E., Nemenoff, R.A. & Das, M. (2002) Hypoxic activation of adventitial fibroblasts: role in vascular remodeling. *Chest* **122**, 326S–334S

42. Naik, J.S., Earley, S., Resta, T.C. & Walker, B.R. (2005) Pressure-induced smooth muscle cell depolarization in pulmonary arteries from control and chronically hypoxic rats does not cause myogenic vasoconstriction. *J. Appl. Physiol.* **98**, 1119–1124
43. Sweeney, M. & Yuan, J.X. (2000) Hypoxic pulmonary vasoconstriction: role of voltage-gated potassium channels. *Respir. Res.* **1**, 40–48
44. Platoshyn, O., Yu, Y., Golovina, V.A., McDaniel, S.S., Krick, S., Li, L., Wang, J.Y., Rubin, L.J. & Yuan, J.X. (2001) Chronic hypoxia decreases K(V) channel expression and function in pulmonary artery myocytes. *Am. J. Physiol. Lung Cell Mol. Physiol.* **280**, L801–L812
45. Ekhterae, D., Platoshyn, O., Krick, S., Yu, Y., McDaniel, S.S. & Yuan, J.X. (2001) Bcl-2 decreases voltage-gated K^+ channel activity and enhances survival in vascular smooth muscle cells. *Am. J. Physiol. Cell Physiol.* **281**, C157–C165
46. Fagan, K.A., Oka, M., Bauer, N.R., Gebb, S.A., Ivy, D.D., Morris, K.G. & McMurtry, I.F. (2004) Attenuation of acute hypoxic pulmonary vasoconstriction and hypoxic pulmonary hypertension in mice by inhibition of Rho-kinase. *Am. J. Physiol. Lung Cell Mol. Physiol.* **287**, L656–L664
47. Hyvelin, J.M., Howell, K., Nichol, A., Costello, C.M., Preston, R.J. & McLoughlin, P. (2005) Inhibition of Rho-kinase attenuates hypoxia-induced angiogenesis in the pulmonary circulation. *Circ. Res.* **97**, 185–191

9

The AMP-activated protein kinase: more than an energy sensor

Louis Hue[1] and Mark H. Rider

Université catholique de Louvain, Christian de Duve Institute of Cellular Pathology, Hormone and Metabolic Research Unit, Avenue Hippocrate, 75, B-1200 Brussels, Belgium

Abstract

The AMPK (AMP-activated protein kinase) is a highly conserved eukaryotic protein serine/threonine kinase. It mediates a nutrient signalling pathway that senses cellular energy status and was appropriately called the fuel gauge of the cell. At the cellular level, AMPK controls energy homoeostasis by switching on catabolic ATP-generating pathways, while switching off anabolic ATP-consuming processes. Its effect on energy balance extends to whole-body energy homoeostasis, because, in the hypothalamus, it integrates nutritional and hormonal signals that control food intake and body weight. The interest in AMPK also stems from the demonstration of its insulin-independent stimulation of glucose transport in skeletal muscle during exercise. Moreover, the potential importance of AMPK in metabolic diseases is supported by the notion that AMPK mediates the anti-diabetic action of biguanides and thiazolidinediones and that it might be involved in the metabolic syndrome. Finally, the more recent demonstration that AMPK activation could occur independently of changes in cellular energy status, suggests that AMPK action extends to the control of non-metabolic functions.

[1]*To whom correspondence should be addressed (email louis.hue@uclouvain.be).*

Introduction

The purpose of this chapter is to summarize the multiple effects of AMPK. Because of space limitations, we have mainly quoted review articles, in which references to the original primary research papers can be found.

AMPK (AMP-activated protein kinase)

AMPK was discovered independently more than 30 years ago as a protein that inactivated ACC (acetyl-CoA carboxylase) and HMG-CoA (hydroxymethylglutaryl-CoA) reductase, two key enzymes involved in lipid metabolism. Almost 15 years later, Hardie's group [1,2] realized that both activities were catalysed by the same multisubstrate protein kinase, which was activated by AMP [1]. This unique feature led to the name AMPK and made it an ideal candidate to sense the adenine nucleotide content of the cell [1,2].

AMPK is a heterotrimeric complex containing a catalytic (α) and two regulatory (β and γ) subunits. Each subunit has multiple isoforms (α1, α2, β1, β2, γ1, γ2, γ3) giving twelve possible combinations of holoenzyme with different tissue distributions. In skeletal muscle, only the α1β2γ1, α2β2γ1 and α2β2γ3 complexes seem to be expressed [3]. The N-terminus of the α-subunit contains a protein kinase domain, with a threonine residue (Thr^{172}) in the activation loop (termed the T-loop), whose phosphorylation by upstream kinases is required for AMPK activation. The β-subunit could act as a scaffold to which the two subunits are bound. It also contains a glycogen-binding domain, whose three-dimensional structure has been determined, but whose physiological role remains to be clarified. The γ-subunit contains four CBS (cystathionine β-synthase) motifs in tandem repeats, which form pairs of so-called Bateman domains that bind AMP. ATP can also bind to these sites, but with much lower affinity. Mutations in the γ2-subunit cause glycogen accumulation, leading to cardiac arrhythmias [1,2].

The intracellular distribution of the two catalytic isoforms of AMPK may differ. When phosphorylated, the α2-subunit translocates to the nucleus during hypoxia or in contracting muscle, whereas the α1-subunit remains cytoplasmic. In specialized tissues such as the carotid body glomus cells, the α1-subunit, which appears as the effector of hypoxic chemotransduction, is targeted to the plasma membrane [4], and in polarized epithelia, AMPK co-localizes with the CFTR (cystic fibrosis transmembrane conductance regulator) protein [5]. The role of the different subunits in AMPK translocation is not known.

Upstream kinases

After many years of intensive research, it was only in 2003 that the first upstream AMPK-kinase was identified as LKB1 [1,6], a tumour suppressor mutated in Peutz–Jeghers syndrome, an autosomal dominant disorder characterized by the development of gastrointestinal polyps and elevated risks of cancer. In the presence of AMP, LKB1 activates AMPK by phosphorylating the Thr^{172} residue of the α-subunit T-loop. AMP does not act directly on LKB1, which is thought to be constitutively active. As a result of its binding to the γ-subunits of AMPK, AMP was proposed to allosterically stimulate AMPK activity and to induce a conformational change exposing Thr^{172} for phosphorylation by LKB1 [7]. The alternative hypothesis is that AMP inhibits Thr^{172} dephosphorylation via a so-called 'substrate-mediated effect' on the inactivating protein phosphatase, PP2C [7,8], whose involvement in the control of AMPK activity in intact cells has been recently demonstrated [9]. Therefore protein phosphatases would be involved in AMPK activation resulting from a rise in AMP. LKB1 is a master kinase, capable of phosphorylating and activating several ARKs (AMPK-related kinases) [10]. The discovery that CaMKKs (calmodulin-dependent protein kinase kinases) could also activate AMPK, but independently of AMP, has broadened the potential roles of AMPK in the control of processes that do not necessarily involve a change in cellular energy status [1,6]. More recently a third AMPK-kinase, the TAK1 (transforming growth factor-β-activated kinase), has been found [11]. TAK1 is part of a G-protein signalling pathway downstream of cytokine receptors, which opens up the possibility that AMPK could be involved in inflammatory responses. Figure 1 summarizes the signalling pathways that activate AMPK.

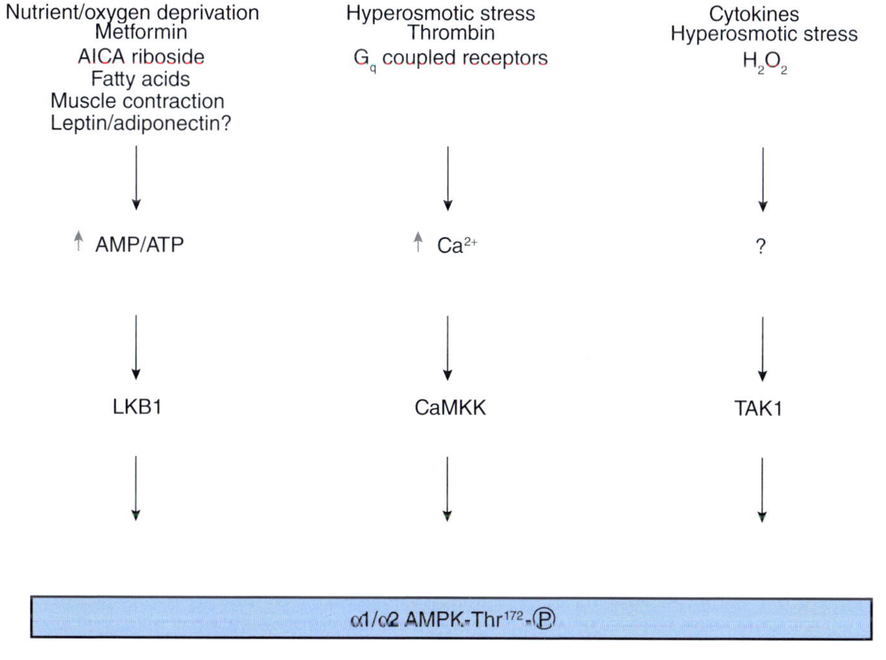

Figure 1. AMPK activation

Various stimuli lead to AMPK activation through different pathways. One pathway senses energy changes and is mediated by LKB1 (Peutz–Jeghers protein), the second results from changes in intracellular calcium concentrations and is mediated by CaMKKs, while the third pathway is mediated by TAK1.

AMPK activation

As stated above, any increase in AMP/ATP ratio results in AMPK activation via phosphorylation of its Thr^{172} by LKB1. Conditions leading to changes in AMP concentration are directly related to changes in ATP concentrations. At any time, intracellular ATP concentrations reflect the balance between energy supply-and-demand. In most cells, ATP supply relies on mitochondrial oxidative phosphorylation, whereas ATP demand depends on the energy required to perform various cell functions. Any decrease in supply (e.g. hypoxia) or increase in demand (e.g. intense exercise) will decrease ATP concentrations. The equilibrium maintained by adenylate kinase translates a fall in ATP to a relatively larger change in AMP, since the AMP/ATP ratio increases as the square of the ADP/ATP ratio, thus providing an amplification mechanism [1].

It is worth noting that the cytosolic concentration of free AMP is less than its total cellular concentration (about 0.2 mM in normoxic tissues). AMP is in fact distributed between the cytosol and mitochondria and binds to several abundant proteins such as glycogen phosphorylase, adenylate kinase and also to many other enzymes. AMP concentrations calculated from ^{31}P-NMR spectroscopy vary from 2 μM in normoxic hearts to about 30 μM in the presence of various inhibitors [12]. These values are at least 10 times lower than the total concentrations but are in the range of AMP concentrations giving half-maximal activation of AMPK (2-50 μM, depending on the concentration of the competitor ATP).

In this section we briefly review the numerous conditions leading to increases in AMP concentration and AMPK activation. A decrease in ATP supply is observed under metabolic stresses, e.g. when oxygen supply is lacking, which in most cases corresponds to non-physiological conditions. This is the case in ischaemic heart where AMPK becomes activated within a few minutes. Similar effects are obtained with mitochondrial inhibitors such as oligomycin [1,13]. Increased ATP demand is another situation leading to AMPK activation, especially when combined with decreased ATP supply. This is the case in contracting muscle, particularly during intense exercise, when the concentration of ATP falls because of the massive energy demand and partial hypoxia resulting from contraction. The extent of AMPK activation seems to depend both on the intensity and duration of exercise [3]. By contrast, AMPK activation does not occur in heart subjected to high workloads and this organ adapts its ATP supply to increased energy demand [14].

In several tissues, AMP and ATP concentrations depend on the availability of glucose, which controls specialized functions. For example, in pancreatic β-cells, ATP concentrations are related to glucose metabolism and insulin secretion [1]. In these cells, AMPK is activated by glucose deprivation, suggesting that AMPK might indirectly control insulin secretion [1]. In certain discrete hypothalamic centres, glucose deprivation also correlates with AMPK activation and with the secretion of various (an)orexigenic peptides that control food intake [2]. Hypothalamic AMPK activation during starvation and its inactivation upon re-feeding are mediated by nutritional and hormonal signals [15]. Indeed, AMPK is inactivated not only by glucose, but also by insulin and leptin [15–17]. Leptin is secreted by adipose tissue and its circulating concentration reflects whole-body fat stores. The systemic concentrations of insulin and leptin decrease during starvation and both inhibit food intake. Over-expression of active forms of AMPK or intra-cerebral injection of AMPK activators stimulates food intake, whereas expression of a dominant-negative mutant of AMPK has the opposite effect. Taken together, these findings support the concept that AMPK participates in the control of whole body energy homoeostasis in the hypothalamus [2,15].

AMP concentrations can increase independently of changes in ATP concentration. The activation of fatty acids to their CoA derivatives by acyl-CoA synthetases produces AMP and PP_i. The rise in AMP and ensuing AMPK activation acts as a nutrient sensor for fatty acids. Under these conditions AMPK activation will favour fatty acid oxidation through the AMPK-mediated inactivation of ACC (see below). This elegant feed-forward mechanism, which has been described in the perfused rat heart [18], ensures that free fatty acids are indeed oxidized rather than esterified.

Several substances and drugs are known to activate AMPK. AICA (5-amino-4-imidazole-carboxamide) riboside is an analogue of adenosine that can be phosphorylated in certain cells to ZMP, which mimics several effects of AMP. Thus AICA riboside has been extensively used in cells or even *in vivo* to activate AMPK [1,2]. However, several effects of AICA riboside, including the inhibition of liver glucose uptake and of mitochondrial respiration, are independent of AMPK [19]. In addition, intracellular ZMP accumulation has been reported to directly modulate the activity of enzymes which bind AMP, such as glycogen phosphorylase and fructose-1,6-bisphosphatase. Clearly results obtained with AICA riboside should be interpreted with caution. Because of its unwanted side-effects, AICA riboside is not well-suited to investigate the consequences of AMPK activation on cell function. Other more specific tools and/or new pharmacological compounds with AMPK selectivity are required. The recent development of new AMPK activators is of obvious interest [20].

Metformin, the most prescribed drug for Type 2 diabetes, restores glucose homoeostasis by increasing glucose disposal and decreasing hepatic glucose production. Metformin activates AMPK and the involvement of AMPK in the anti-diabetic effect of this drug is supported by studies in LKB1-deficient mice where the blood glucose lowering effect of the compound was greatly decreased [21,22]. Metformin was initially reported to activate AMPK without affecting adenine nucleotide concentrations. This interpretation had to be revised, because the initial effect of metformin is to partially inhibit respiratory chain complex I, thus decreasing ATP production. The inhibition is readily observed at high concentrations (>1 mM), but is already present at low (0.1 mM) concentrations corresponding to the therapeutic range in the portal vein [23]. Thiazolidinediones, another class of anti-diabetic compounds, activate AMPK through an increase in AMP/ATP ratio, which also seems to result from an inhibition of mitochondrial respiration [22]. In relation to the effect of these anti-diabetic drugs, AMPK deficiency has been proposed to underlie the progression of the metabolic syndrome, which is characterized by insulin resistance [24]. This syndrome corresponds to a combination of various metabolic and haemodynamic factors that represent a high risk of developing Type 2 diabetes and cardiovascular disease. Hyperglycaemia, hypertension, abdominal obesity, low HDL (high-density lipoprotein) cholesterol and high circulating triacylglycerols are characteristic features. Interestingly, mice that are deficient in the AMPK α2-subunit display abnormal control of body mass, insulin sensitivity, glucose and lipid homoeostasis and mimic the metabolic syndrome to some extent [21]. In addition, the circulating levels of the fat-cell derived adiponectin are decreased in obese and insulin-resistant patients and its insulin-sensitizing effect might well be mediated by AMPK activation [2,15,21,25].

Several conditions are known to change AMPK activity without detectable changes in adenine nucleotide concentrations. The effect of insulin to antagonize hypoxia-induced AMPK activation is not related to changes in adenine nucleotides but could result from phosphorylation by PKB (protein kinase B) of $Ser^{485/491}$ on the α-subunits of AMPK, which inhibits Thr^{172} phosphorylation by LKB1 [26] (see below). Since CaMKKβ has been reported to be an activating upstream AMPK kinase, AMPK activation could also be triggered by an increase in cytosolic calcium. This could explain AMPK activation by thrombin in endothelial cells [27] and in response to triggering of the T-cell antigen receptor in lymphocytes [28]. It could also apply to AMPK activation by osmotic stress, which occurs without a change in adenine nucleotide concentrations [29]. Finally, it is worth mentioning that, by contrast with its inactivating effect on hypothalamic AMPK, leptin has been reported to activate AMPK in muscle, as does adiponectin [2,15,21,25]. The precise mechanism of AMPK (in)activation by these adipokines is controversial.

AMPK targets

Key enzymes involved in the control of carbohydrate, lipid and protein metabolism have been identified as AMPK substrates, leading to the concept that AMPK acts as a metabolic master switch that conserves ATP. A list of some of the targets of AMPK and their phosphorylation site sequences is given in Table 1. The substrate recognition motif for AMPK is ϕ(X,β)XXS/TXXXϕ, where ϕ is a hydrophobic residue, β a basic residue, X is any amino acid and the parentheses indicate that the order of residues at the P-4 and P-3 positions is not critical. Table 1 shows that AMPK is predominantly a serine-directed protein kinase and that in some targets the consensus is not perfectly adhered to. One interesting possibility is that the various AMPK isoforms might differ in substrate recognition. In the hearts of mice lacking the α2 catalytic subunits,

Table 1. Sites phosphorylated by AMPK

The phosphorylated serine/threonine residue are indicated by underlining, hydrophobic residues that form the AMPK consensus are in bold and basic residues of the consensus are in italics.

Phosphorylation site in AMPK substrate	Phosphorylation site sequence	Comments A: activation; I: inactivation	Reference
Metabolic targets			
Ser^{79} of rat acetyl-CoA carboxylase (ACC1)	H**M***R*SSMSGLH**L**V*K*	**I.** Inhibitory for fatty acid synthesis	[30]
Ser^{221} of human acetyl-CoA carboxylase (ACC2)	T**M***R*PSMSGLH**L**V*K*	**I/A.** Inhibits ACC2 but stimulates fatty acid oxidation by lowering [malonyl-CoA]. Site identified by homology with the site in rat ACC1	
Ser^{871} of rat hydroxymethylglutaryl-CoA reductase (HMG-CoA reductase)	H**M**V*H*NRSKIN**L**QD	**I.** Inhibitory for cholesterol biosynthesis	[31]
Ser^{565} of rat hormone-sensitive lipase (HSL)	S**M***R*RSVSEAA**L**AQ	**I.** Ser^{563} upstream of Ser^{565} is phosphorylated by PKA	[32]
Ser^{7} of rabbit muscle glycogen synthase (GS)	P**L***S*RTLSVSS**L**PG	**I.** This is the so-called 'site 2' that is also phosphorylated by PKA	[33]
Ser^{466} of bovine heart 6-phosphofructo-2-kinase (PFK2-B2)	VRM*R*RNSFTP**L**SS	**A.** This site can be phosphorylated by PKA, PKCs and insulin-stimulated protein kinases such as PKB	[34]
Ser^{461} of human inducible 6-phosphofructo-2-kinase (PFK2FB3)	P**L**M*R*RNSVTP**L**AS	**A.** Stimulatory for glycolysis in activated monocytes subjected to hypoxia	[35]
Transcription factors			
Ser^{171} of rat transducer of regulated CREB activity (TORC2)	A**L**N*R*TSSDSA**L**HT	**I.** This site is also phosphorylated by the AMPK-related SIK2 (salt-inducible kinase-2). Phosphorylation sequesters TORC in the cytoplasm where it cannot co-activate CREB	[36]
Ser^{568} of rat carbohydrate-response element-binding protein (ChREBP)	L**L***R*PPESPDA**V**PE	**I.** Decreases DNA binding activity to the promoter of the L-type pyruvate kinase gene	[37]

(*continued*)

Table 1. Sites phosphorylated by AMPK (*continued*)

The phosphorylated serine/threonine residue are indicated by underlining, hydrophobic residues that form the AMPK consensus are in bold and basic residues of the consensus are in italics.

Phosphorylation site in AMPK substrate	Phosphorylation site sequence	Comments A: activation; I: inactivation	Reference
Ser^{304} of human hepatic nuclear factor 4α (HNF4α)	K**I**K*R*L*R*SQVQ**V**SL	**I.** Decreases formation of homodimers and DNA binding, and promotes degradation	[38]
Ser^{89} of human transcriptional co-activator p300	E**L**L*R*SGSSPN**L**NM	**I.** Blocks its ability to interact with nuclear receptors such as PPARγ	[39]
Proteins involved in the control of translation/cell growth			
Thr^{2446} of the human mammalian target of rapamycin (mTOR)	KRS*R*TRTDSYSAG	**I.** Ser^{2448} downstream of this site can be phosphorylated in response to insulin by PKB or p70S6K	[40]
Thr^{1227} of rat tuberous sclerosis complex 2 (TSC2)	T**L**P*R*SNTVAS**F**SS	**A.** This, or the next site, may stabilize the TSC1–TSC2 complex	[41]
Ser^{1345} of rat tuberous sclerosis complex 2 (TSC2)	P**L**S*K*SSSSPE**L**QT	**A.** This, or the previous site, may stabilize the TSC1–TSC2 complex	[41]
Ser^{398} of human eukaryotic elongation factor 2 (eEF2)-kinase	S**L**PSSPSSATPHS	**A.** Inhibits protein synthesis elongation by increasing eEF2 Thr^{56} phosphorylation	[42]
Signalling proteins			
Thr^{494} of human endothelial NO synthase (NOS III)	G**I**T*R*KKTFKE**V**AN	**A.** Increases activity at low [Ca^{2+}/calmodulin]	[43]
Ser^{1176} of human endothelial NO synthase (NOS III)	SRI*R*TQSFSLQER	**A.** This site can be phosphorylated by PKB	[43]
Ser^{789} of rat insulin receptor substrate-1 (IRS-1)	H**L**R*L*SSSSGR**L**RY	**A.** This site is also phosphorylated by QIK (qin-induced kinase, also known as SIK2).	[44]
Ser^{621} of human/rat Raf-1	K**I**N*R*SASEPS**L**HR	**I?** The role of this phosphorylation in Raf-1 function is unclear	[45]

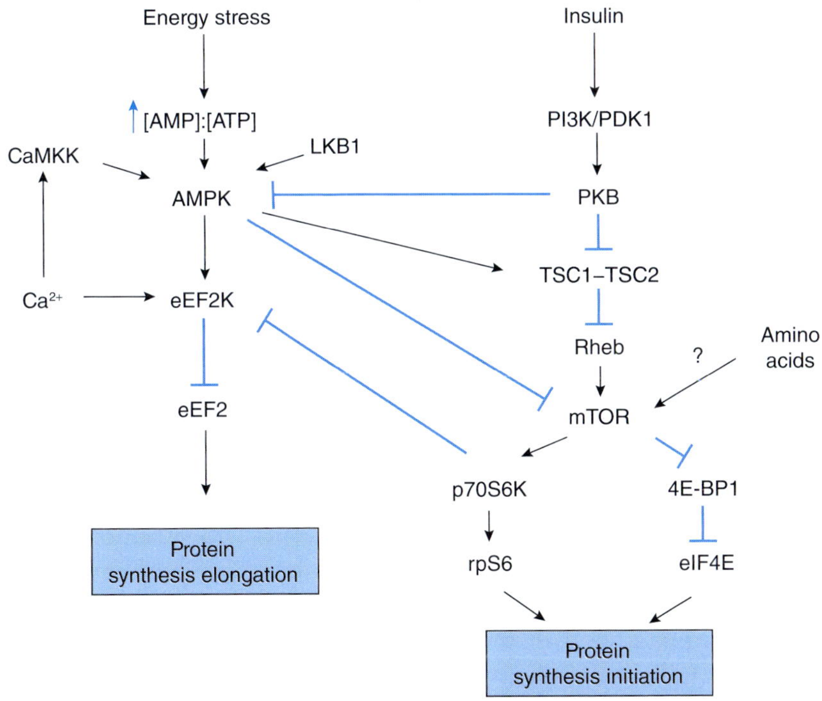

Figure 2. Antagonistic effects of AMPK and insulin signalling on protein synthesis
AMPK is activated in response to energy stress via a rise in the AMP/ATP ratio (LKB1 pathway) and potentially in situations where intracellular calcium concentrations increase (CaMKK pathway). Once activated, AMPK inhibits protein synthesis via a reduction in mTOR signalling by increasing eEF2 phosphorylation. Insulin not only stimulates protein synthesis by stimulating the PKB/mTOR pathway, but also antagonizes AMPK signalling by counteracting AMPK activation. In the scheme, the arrows (black) indicate steps leading to stimulation or activation, whereas bars with a cross-head (blue) indicate steps of inhibition or inactivation. PI3K, phosphoinositide 3-kinase; PDK1, phosphoinositide dependent kinase 1; rpS6 (ribosomal protein S6).

the metabolic response to no-flow ischaemia is abrogated, indicating that the persisting α1-subunit cannot compensate for the lack of the α2 catalytic subunit [46].

A number of studies have shown that AMPK activation inhibits fatty acid and cholesterol synthesis in liver and glycogen synthesis in muscle through the phosphorylation and inactivation of ACC, HMGCoA reductase and glycogen synthase respectively. The long-term induction of gluconeogenic enzymes by starvation is also inhibited by AMPK, which phosphorylates TORC2 (transducer of regulated CREB activity 2), a transcriptional co-activator that inhibits the expression of these genes [21,22].

AMPK activation also inhibits protein synthesis at several levels (summarized in Figure 2). Control of protein synthesis is exerted via the (de)phosphorylation of various translation factors and ribosomal proteins [21,22,47], including the mTOR (mammalian-target-of-rapamycin). mTOR activation involves the G protein Rheb (Ras homologue enriched in brain) and the complex TSC1–TSC2 (Tuberous Sclerosis Complex 1 and 2, also called hamartin and tuberin respectively). TSC1–TSC2 is mutated in tuberous sclerosis, a genetic disorder characterized by benign tumours (hamartomas), and acts as a GTPase-activating protein favouring the GDP-bound state of Rheb, thus inhibiting mTOR signalling. PKB, which mediates insulin effects on protein synthesis, phosphorylates and inactivates TSC2 thus keeping Rheb in its active form and favouring mTOR activation. AMPK has the opposite effect by phosphorylating TSC2 on sites other than those phosphorylated by PKB, thereby stimulating GTPase activity and keeping Rheb and mTOR inactive [41]. The mTOR pathway controls initiation via the phosphorylation of 4E-BP1 (eukaryotic initiation factor 4E-binding protein-1), thereby relieving its inhibition on eIF4E, which can then bind the mRNA cap. mTOR also controls elongation via a phosphorylation cascade involving p70S6K (p70 ribosomal protein S6 kinase). p70S6K phosphorylates and inactivates eEF2 (eukaryotic elongation factor 2)-kinase, a dedicated Ca^{2+}-calmodulin-dependent kinase that phosphorylates and inactivates eEF2. Therefore mTOR activation decreases eEF2 phosphorylation, a mechanism involved in the insulin-induced stimulation of protein synthesis [47]. By contrast, AMPK activates eEF2-kinase, which in turn phosphorylates and inactivates eEF2, thus inhibiting translation elongation [21,48]. Taken together, these observations support the concept that AMPK activation inhibits anabolic ATP-consuming processes including protein synthesis and cell growth (Figure 2).

Other studies have demonstrated that AMPK activation favours ATP-producing pathways. Inactivation of ACC by AMPK decreases the concentration of its product malonyl-CoA, thereby relieving inhibition of the carnitine-mediated import of long chain acyl-CoAs into mitochondria and eventually stimulating their oxidation [1,21,22]. This control is probably the reason for the presence of ACC in non-lipogenic tissues. Fatty acid oxidation requires oxygen and obviously cannot be stimulated in ischaemic tissues, i.e. when AMPK is activated. However, the beneficial effect of AMPK activation on fatty acid oxidation is expected during heart re-perfusion following a severe ischaemic episode. AMPK activation also stimulates ATP production by increasing glycolytic flux at two levels. In exercising muscle, AMPK activation stimulates glucose uptake by increasing the recruitment of GLUT4 (glucose transporter 4) transporters to the plasma membrane in an insulin-independent manner. This underlies the recommendation of physical exercise for diabetic patients. The molecular mechanism could involve the phosphorylation by AMPK of the Rab GTPase activating protein, AS160 (Akt substrate 160), a target of PKB in the insulin-signalling pathway [49]. By analogy with the control of mTOR by Rheb, the GTPase activity of AS160 is inhibited via phosphorylation by PKB, thereby maintaining Rabs that regulate GLUT4 translocation in their active GTP-bound state [49]. In hypoxic heart and monocytes, AMPK also phosphorylates and activates 6-phosphofructo-2-kinase, thereby increasing the concentration of its product fructose 2,6-bisphosphate [13,35]. Thus protein phosphorylation seems to be part of the mechanism by which oxygen deprivation stimulates glycolysis (the Pasteur effect). The combined increase in fructose 2,6-bisphosphate and AMP, both of which allosterically stimulate 6-phosphofructo-1-kinase activity, contributes to the overall stimulation of glycolysis. Whether AMPK also influences ATP production by mitochondria does not seem to be the case, at least in the short-term. However, AMPK activation stimulates mitochondrial biogenesis, probably through up-regulation of PGC-1α (peroxisome-proliferator-activated receptor γ co-activator-1α) [50].

The HIF-1 (hypoxia-inducible factor) complex is another oxygen sensor allowing cells to adapt to oxygen deprivation. This transcription factor stimulates the expression of several genes by binding to hypoxia-responsive elements in promoters of key angiogenic and glycolytic genes. The role of AMPK in the regulation of HIF-1 is unclear. Paradoxically, AMPK should inhibit HIF-1 synthesis, by inhibiting the mTOR pathway, which seems to control its synthesis. However, the available evidence suggests that hypoxia independently activates both AMPK and HIF-1 [51], implying that activation of pre-existing HIF-1 would be enough to trigger the adaptive response to low oxygen. Therefore both AMPK activation and HIF-1 stabilization appear to be parallel responses to oxygen deprivation, which act synergistically to maintain energy homeostasis. Another interesting target is eNOS (endothelial nitric oxide synthase), which can be phosphorylated and activated by AMPK, thereby contributing to vasodilation and improved oxygen supply [43].

In growing cells, protein synthesis and ion transport together are the most prominent consumers of oxidatively-derived ATP. Since protein synthesis is inhibited by AMPK activation at several levels, one would expect energy-consuming ion transport systems, such as Na^+/K^+-ATPase, Ca^{2+}-ATPase or other transporters to be shut down as a result of AMPK activation. Accordingly, there is some evidence that several ion channels could be regulated by AMPK [5,52].

Some caution needs to be exercised before a target protein can be considered as a *bona fide* AMPK substrate. For example, many AMPK targets have been reported on the basis of over-expression of recombinant proteins in cells incubated with pharmacological activators of AMPK, such as AICA riboside, or subjected to treatments that deplete intracellular ATP, conditions which are somewhat artificial. Also in some cases, the sites for AMPK have not been identified formally by direct sequencing or mass spectrometry, but by searching target protein sequences for the AMPK consensus and then mutating the putative phosphorylation site to a negatively charged (to mimic phosphorylation) or non-phosphorylatable residue and looking at effects on function. To verify that AMPK is the physiologically relevant kinase for a particular target, changes in cellular AMPK activity should be correlated with the extent of phosphorylation of the substrate protein at the relevant site. This could be undertaken either by incubating cells with a selective AMPK inhibitor/activator or by modulating AMPK content by genetic approaches. However, these approaches might be inadequate in one way or another, for example either by lack of isoform specificity with the dominant negative approach, by incomplete AMPK inhibition or by compensatory mechanisms such as induction of the expression of the other AMPK catalytic subunit isoform or indeed of other related kinases. The overexpression of a dominant-negative construct might not target endogenous AMPK in the right intracellular compartment or might titrate out signalling molecules necessary for the activation of other kinase cascades that converge on AMPK targets. Clearly, where possible several approaches should be tested in controlled experiments where AMPK activity is monitored in cells, for example by measuring ACC phosphorylation.

Sharing control and cross-talks

Virtually all the known AMPK targets are multi-site phosphorylated proteins either containing more than one AMPK site, or sites for other protein kinases such as PKA (cyclic AMP-dependent protein kinase) and PKB. AMPK phosphorylation sites are often adjacent to sites phosphorylated by other protein kinases, or the AMPK site itself is also phosphorylated by a protein kinase from another signalling pathway (Table 1). There are cases where the AMPK and insulin signalling pathways (via PKB) converge on the same target at the same site and with the same effects, as e.g. heart PFK-2 (6-phosphofructo-2-kinase) [34] and skeletal muscle GLUT4 transporters [53].

By contrast, the insulin and AMPK pathways have opposite effects on protein synthesis (Figure 2). Insulin stimulates protein synthesis by activating the mTOR pathway [47], whereas AMPK activation decreases protein synthesis via eEF2-kinase activation and by inhibiting mTOR signalling [21,54], but in some cases the latter effect was only evident after the mTOR pathway had first been stimulated by growth factors/insulin or amino acids [14,55]. PKB and AMPK have been reported to phosphorylate distinct sites both on TSC2 [41] and mTOR [40,56] with opposing effects. Therefore the possibility of hierarchical phosphorylation of these proteins by PKB and AMPK obviously needs investigation. Indeed the effect of insulin to antagonize AMPK activation involves a hierarchical mechanism whereby phosphorylation of the α-subunit $Ser^{485/491}$ by PKB reduces subsequent phosphorylation of Thr^{172} by LKB1 and the resulting AMPK activation [26].

Anti-proliferative and non-metabolic effects of AMPK

Although LKB1 seems to be the major upstream activating kinase in many tissues and cells, the fact that signalling via CaMKKs can lead to AMPK activation indicates that both changes in intracellular calcium and AMP concentrations may play separate or inter-dependent roles in the regulation of AMPK activity. Moreover, it was recently shown that TAK1 was an authentic AMPK kinase at least *in vitro* [11]. Therefore although the primary targets for AMPK have previously been presumed to be mainly involved in energy homoeostasis, AMPK activation seems now to control non-metabolic processes, such as cell growth, progression through the cell cycle and the organization of the cytoskeleton.

Sustained AMPK activation suppresses cell proliferation and growth and can even trigger apoptosis [51,57]. This anti-proliferative action of AMPK is mediated by a number of tumour suppressors that are part of the AMPK signalling pathway, such as LKB1 and TSC1–TSC2 [58]. In addition, AMPK activation in epithelial cells has also been shown to control tight junction formation [59] and cell migration through re-organization of the actin cytoskeleton [60]. Finally, thrombin has been shown to activate AMPK in endothelial cells through a calcium-dependent mechanism mediated by CaMKKβ [27]. It is clear that these new unexpected roles of AMPK extend beyond energy homoeostasis and potentially link AMPK to the many calcium-dependent cell functions.

Conclusion

Early knowledge about the AMPK system led to an understanding of its role as an energy-sensor. Accumulating evidence now indicates that AMPK can be considered as a multi-tasking protein controlling cellular homoeostasis at large. This might be related to the fact that it is encoded by house-keeping genes. One might speculate that the two catalytic subunits of AMPK have different specificities, the α2-subunit being more metabolically orientated and controlled by LKB1, whereas the α1-subunit of AMPK would be involved in various calcium-dependent cell functions and controlled by CaMKKs. Which catalytic subunit is devoted to the control of the cell cycle needs further investigation.

The number of AMPK substrates is still growing, as is the number of papers published on AMPK. However, the demonstration that reported target proteins can be considered as *bona fide* substrates is often lacking. Many more targets are likely to be identified in view of the fact that AMPK is an ancient eukaryotic kinase and that its role extends beyond that of controlling energy homoeostasis.

Summary

- *AMPK is a heterotrimeric protein serine/threonine kinase containing one catalytic (α) and two (β, γ) regulatory subunits, which is activated by phosphorylation of its α-subunit T-loop Thr^{172} residue by upstream kinases*
- *Three pathways lead to AMPK activation: one sensing energy depletion and mediated by LKB1; the second resulting from changes in intracellular calcium concentrations and mediated by CaMKKs; and the third mediated by TAK1.*
- *AMPK controls energy homoeostasis: at the cellular level, by phosphorylating key metabolic enzymes resulting in the stimulation of ATP production and inhibition of ATP consumption; at the whole-body level, by controlling food intake in the hypothalamus.*

- *AMPK activation also controls non-metabolic processes, such as cell growth, progression through the cell cycle and the organization of the cytoskeleton, by mechanisms that are not necessarily linked to energy homoeostasis.*

The work carried out in the authors' laboratory was supported by the Belgian Fund for Medical Scientific Research, the Interuniversity Poles of Attraction Belgian Science Policy (P5/05), the French Community of Belgium (Actions de Recherche Concertées) and the EXGENESIS Integrated Project (LSHM-CT-2004-005272) from the European Commission.

References

1. Hardie, D.G., Hawley, S.A. & Scott, J.W (2006) AMP-activated protein kinase: development of the energy sensor concept. *J. Physiol.* **574**, 7–15
2. Kahn, B.B., Alquier, T., Carling, D. & Hardie, D.G. (2005) AMP-activated protein kinase: ancient energy gauge provides clues to modern understanding of metabolism. *Cell Metab.* **1**, 15–25
3. Jørgensen, S.B., Richter, E.A. & Wojtaszewski, J.F.P. (2006) Role of AMPK in skeletal muscle metabolic regulation and adaptation in relation to exercise. *J. Physiol.* **574**, 17–31
4. Wyatt, C.N., Mustard, K.J., Pearson, S.A., Dallas, M.L., Atkinson, L., Kumar, P., Peers, C., Hardie, D.G. & Evans, A.M. (2007) AMP-activated protein kinase mediates carotid body excitation by hypoxia. *J. Biol. Chem.* **282**, 8092–8098
5. Hallows, K.R., Raghuram, V., Kemp, B.E., Witters, L.A. & Foskett, J.K. (2000) Inhibition of cystic fibrosis transmembrane conductance regulator by novel interaction with the metabolic sensor AMP-activated protein kinase. *J. Clin. Invest.* **105**, 1711–1721
6. Witters, L.A., Kemp, B.E. & Means, A.R. (2006) Chutes and ladders: the search for protein kinases that act on AMPK. *Trends Biochem. Sci.* **31**, 13–16
7. Davies, S.P., Helps, N.R., Cohen, P.W. & Hardie D.G. (1995) 5′-AMP inhibits dephosphorylation, as well as promoting phosphorylation, of the AMP-activated protein kinase. Studies using bacterially expressed human protein phosphatase 2C-α and native bovine protein phosphatase-2Ac. *FEBS Lett.* **377**, 421–425
8. Sanders, M.J., Grondin, P.O., Hegarty, B.D., Snowden, M.A. & Carling D. (2007) Investigating the mechanism for AMP activation of the AMP-activated protein kinase cascade. *Biochem. J.* **403**, 139–148
9. Steinberg, G.R., Michell, B.J., van Denderen, B.J., Watt, M.J., Carey, A.L., Fam, B.C., Andrikopoulos, S., Proietto, J., Gorgin, C.Z., Carling, D. et al. (2006) Tumor necrosis factor α-induced skeletal muscle insulin resistance involves suppression of AMP-kinase signalling. *Cell Metab.* **4**, 465–474
10. Lizcano, J.M., Goransson, O., Toth, R., Deak, M., Morrice, N.A., Boudeau, J., Hawley, S.A., Udd, L., Makela, T.P., Hardie, D.G. & Alessi, D.R. (2004) LKB1 is a master kinase that activates 13 kinases of the AMPK subfamily, including MARK/PAR-1. *EMBO J.* **23**, 833–843
11. Momcilovic, M., Hong, S.P. & Carlson, M. (2006) Mammalian TAK1 activates SNF1 protein kinase in yeast and phosphorylates AMP-activated protein kinase *in vitro*. *J. Biol. Chem.* **281**, 25336–25343
12. Frederich, M. & Balschi, J.A. (2002) The relationship between AMP-activated protein kinase activity and AMP concentration in the isolated perfused rat heart. *J. Biol. Chem.* **277**, 1928–1932
13. Marsin, A.S., Bertrand, L., Rider, M.H., Deprez, J., Beauloye, C., Vincent, M.F., Van den Berghe, G., Carling, D. & Hue, L. (2000) Phosphorylation and activation of heart PFK-2 by AMPK has a role in the stimulation of glycolysis during ischaemia. *Curr. Biol.* **10**, 1247–1255
14. Horman, S., Beauloye, C., Vertommen, D., Vanoverschelde, J.L., Hue, L. & Rider, M.H. (2003) Myocardial ischemia and increased heart work modulate the phosphorylation state of eukaryotic elongation factor-2. *J. Biol. Chem.* **278**, 41970–41976

15. Xue, B. & Kahn, B. (2006) AMPK integrates nutrient and hormonal signals to regulate food intake and energy balance through effects in the hypothalamus and peripheral tissues. *J. Physiol.* **574**, 73–83
16. Mountjoy, P.D., Bailey, S.J. & Rutter, G.A. (2007) Inhibition by glucose or leptin of hypothalamic neurons expressing neuropeptide Y requires changes in AMP-activated protein kinase activity. *Diabetologia* **50**, 168–177
17. Cai, F., Gyulkhandanyan, A.V., Wheeler, M.B. & Belsham, D.D. (2007) Glucose regulates AMP-activated protein kinase activity and gene expression in clonal, hypothalamic neurons expressing proopiomelanocortin: additive effects of leptin or insulin. *J. Endocrinol.* **192**, 605–614
18. Clark, H., Carling, D. & Saggerson, D. (2004) Covalent activation of heart AMP-activated protein kinase in response to physiological concentrations of long-chain fatty acids. *Eur. J. Biochem.* **271**, 2215–2224
19. Guigas, B., Bertrand, L., Taleux, N., Foretz, M., Wiernsperger, N., Vertommen, D., Andreelli, F., Viollet, B. & Hue, L. (2006) 5-aminoimidazole-4-carboxamide-1-β-D-ribofuranoside and metformin inhibit hepatic glucose phosphorylation by an AMP-activated protein kinase-independent effect on glucokinase translocation. *Diabetes* **55**, 865–874
20. Cool, B., Zinker, B., Chiou, W., Kifle, L., Cao, N., Perham, M., Dickinson, R., Adler, A., Gagne, G., Iyengar, R. et al. (2006) Identification and characterization of a small molecule AMPK activator that treats key components of type 2 diabetes and the metabolic syndrome. *Cell. Metab.* **3**, 403–416
21. Viollet, B., Foretz, M., Guigas, B., Horman, S., Dentin, R., Bertrand, L., Hue, L. & Andreelli, F. (2006) Activation of AMP-activated protein kinase in the liver: a new strategy for the management of metabolic hepatic disorders. *J. Physiol.* **574**, 41–53
22. Hardie, D.G. (2007) AMP-activated protein kinase as a drug target. *Annu. Rev. Pharmacol. Toxicol.* **47**, 185–210
23. Guigas, B., Detaille, D., Chauvin, C., Batandier, C., De Oliveira, F., Fontaine, E. & Leverve, X. (2004) Metformin inhibits mitochondrial permeability transition and cell death: a pharmacological *In vitro* study. *Biochem. J.* **382**, 877–884
24. Ruderman, N.B. & Saha, A.K. (2006) Metabolic syndrome: adenosine monophosphate-activated protein kinase and malonyl coenzyme A. *Obesity* **14** (Suppl 1) 25S–33S
25. Daval, M., Foufelle, F. & Ferré, P. (2006) Functions of AMP-activated protein kinase in adipose tissue. *J. Physiol.* **574**, 55–62
26. Horman, S., Vertommen, D., Heath, R., Neumann, D., Mouton, V., Woods, A., Schlattner, U., Wallimann, T., Carling, D., Hue, L. & Rider, M.H. (2006) Insulin antagonizes ischemia-induced Thr172 phosphorylation of AMP-activated protein kinase α-subunits in heart via hierarchical phosphorylation of Ser485/491. *J. Biol. Chem.* **281**, 5335–5340
27. Stahmann, N., Woods, A., Carling, D. & Heller, R. (2006) Thrombin activates AMP-activated protein kinase in endothelial cells via a pathway involving Ca^{2+}/calmodulin-dependent protein kinase kinase β. *Mol. Cell. Biol.* **26**, 5933–5945
28. Tamas, P., Hawley, S.A., Clarke, R.G., Mustard, K.J., Green, K., Hardie, D.G. & Cantrell, D.A. (2006) Regulation of the energy sensor AMP-activated protein kinase by antigen receptor and Ca^{2+} in T lymphocytes. *J. Exp. Med.* **203**, 1665–1670
29. Woods, A., Dickerson, K., Heath, R., Hong, S.P., Momcilovic, M., Johnstone, S.R., Carlson, M. & Carling, D. (2005) Ca^{2+}/calmodulin-dependent protein kinase kinase-β acts upstream of AMP-activated protein kinase in mammalian cells. *Cell Metab.* **2**, 21–33
30. Munday, M.R., Campbell, D.G., Carling, D. & Hardie, D.G. (1988) Identification by amino acid sequencing of three major regulatory phosphorylation sites on rat acetyl-CoA carboxylase. *Eur. J. Biochem.* **175**, 331–338
31. Ching, Y.P., Davies, S.P. & Hardie, D.G. (1996) Analysis of the specificity of the AMP-activated protein kinase by site-directed mutagenesis of bacterially expressed 3-hydroxy 3-methylglutaryl-CoA reductase, using a single primer variant of the unique-site-elimination method. *Eur. J. Biochem.* **237**, 800–808
32. Garton, A.J., Campbell, D.G., Carling, D., Hardie, D.G., Colbran, R.J. & Yeaman, S.J. (1989) Phosphorylation of bovine hormone-sensitive lipase by the AMP-activated protein kinase. A possible antilipolytic mechanism. *Eur. J. Biochem.* **179**, 249–254

33. Carling, D. & Hardie, D.G. (1989) The substrate and sequence specificity of the AMP-activated protein kinase. Phosphorylation of glycogen synthase and phosphorylase kinase. *Biochim. Biophys. Acta* **1012**, 81–86
34. Rider, M.H., Bertrand, L., Vertommen, D., Michels, P.A., Rousseau, G.G. & Hue, L. (2004) 6-Phosphofructo-2-kinase/fructose-2,6-bisphosphatase: head-to-head with a bifunctional enzyme that controls glycolysis. *Biochem. J.* **381**, 561–579
35. Marsin, A.S., Bouzin, C., Bertrand, L. & Hue, L. (2002) The stimulation of glycolysis by hypoxia in activated monocytes is mediated by AMP-activated protein kinase and inducible 6-phosphofructo-2-kinase. *J. Biol. Chem.* **277**, 30778–30783
36. Koo, S.H., Flechner, L., Qi, L., Zhang, X., Screaton, R.A., Jeffries, S., Hedrick, S., Xu, W., Boussouar, F., Brindle, P., Takemori, H. & Montminy, M. (2005) The CREB coactivator TORC2 is a key regulator of fasting glucose metabolism. *Nature* **437**, 1109–1111
37. Kawaguchi, T., Osatomi, K., Yamashita, H., Kabashima, T. & Uyeda, K. (2002) Mechanism for fatty acid 'sparing' effect on glucose-induced transcription: regulation of carbohydrate-responsive element-binding protein by AMP-activated protein kinase. *J. Biol. Chem.* **277**, 3829–3835
38. Hong, Y.H., Varanasi, U.S., Yang, W. & Leff, T. (2003) AMP-activated protein kinase regulates HNF4-α transcriptional activity by inhibiting dimer formation and decreasing protein stability. *J. Biol. Chem.* **278**, 27495–27501
39. Yang, W., Hong, Y.H., Shen, X.Q., Frankowski, C., Camp, H.S. & Leff, T. (2001) Regulation of transcription by AMP-activated protein kinase: phosphorylation of p300 blocks its interaction with nuclear receptors. *J. Biol. Chem.* **276**, 38341–38344
40. Cheng, S.W., Fryer, L.G., Carling, D. & Shepherd, P.R. (2004) Thr2446 is a novel mammalian target of rapamycin (mTOR) phosphorylation site regulated by nutrient status. *J. Biol. Chem.* **279**, 15719–15722
41. Inoki, K., Zhu, T. & Guan, K.L. (2003) TSC2 mediates cellular energy response to control cell growth and survival. *Cell* **115**, 577–590
42. Browne, G.J., Finn, S.G. & Proud, C.G. (2004) Stimulation of the AMP-activated protein kinase leads to activation of eukaryotic elongation factor 2 kinase and to its phosphorylation at a novel site, serine 398. *J. Biol. Chem.* **279**, 12220–12231
43. Chen, Z.P., Mitchelhill, K.I., Michell, B.J., Stapleton, D., Rodriguez-Crespo, I., Witters, L.A., Power, D.A., Ortiz de Montellano, P.R. & Kemp, B.E. (1999) AMP-activated protein kinase phosphorylation of endothelial NO synthase. *FEBS Lett* **443**, 285–289
44. Horike, N., Takemori, H., Katoh, Y., Doi, J., Min, L., Asano, T., Sun, X.J., Yamamoto, H., Kasayama, S., Muraoka, M., Nonaka, Y. & Okamoto, M. (2003) Adipose-specific expression, phosphorylation of Ser794 in insulin receptor substrate-1, and activation in diabetic animals of salt-inducible kinase-2. *J. Biol. Chem.* **278**, 18440–18447
45. Sprenkle, A.B., Davies, S.P., Carling, D., Hardie, D.G. & Sturgill, T.W. (1997) Identification of Raf-1 Ser621 kinase activity from NIH 3T3 cells as AMP-activated protein kinase. *FEBS Lett.* **403**, 254–258
46. Zarrinpashneh, E., Carjaval, K., Beauloye, C., Ginion, A., Mateo, P., Pouleur, A.C., Horman, S., Vaulont, S., Hoerter, J., Viollet, B. et al. (2006) Role of the α2-isoform of AMP-activated protein kinase in the metabolic response of the heart to no-flow ischemia. *Am. J. Physiol. Heart Circ. Physiol.* **291**, H2875–H2883
47. Proud, C.G. (2007) Signalling to translation: how signal transduction pathways control the protein synthetic machinery. *Biochem. J.* **403**, 217–234
48. Horman, S., Browne, G, Krause, U, Patel, J., Vertommen, D., Bertrand, L., Lavoinne, A., Hue, L., Proud, C. & Rider, M. (2002) Activation of AMP-activated protein kinase leads to phosphorylation of elongation factor 2 and an inhibition of protein synthesis. *Curr. Biol.* **12**, 1419–1423
49. Watson, R.T. & Pessin, J.E. (2006) Bridging the gap between insulin signaling and GLUT4 translocation. *Trends Biochem. Sci.* **31**, 215–222
50. Reznick R.M. & Shulman, G.I. (2006) The role of AMP-activated protein kinase in mitochondrial biogenesis. *J. Physiol.* **574**, 33–39
51. Shaw, R.J. (2006) Glucose metabolism and cancer. *Curr. Opin. Cell. Biol.* **18**, 598–608

52. Sukhodub, A., Jovanovic, S., Du, Q., Budas, G., Clelland, A.K., Shen, M., Sakamoto, K., Tian, R. & Jovanovic, A. (2007) AMP-activated protein kinase mediates preconditioning in cardiomyocytes by regulating activity and trafficking of sarcolemmal ATP-sensitive K(+) channels. *J. Cell Physiol.* **210**, 224–236
53. Kramer, H.F., Witczak, C.A., Fujii, N., Jessen, N., Taylor, E.B., Arnolds, D.E., Sakamoto, K., Hirshman, M.F. & Goodyear, L.J. (2006) Distinct signals regulate AS160 phosphorylation in response to insulin, AICAR, and contraction in mouse skeletal muscle. *Diabetes* **55**, 2067–2076
54. Bolster, D.R., Crozier, S.J., Kimball, S.R. & Jefferson, L.S. (2002) AMP-activated protein kinase suppresses protein synthesis in rat skeletal muscle through down-regulated mammalian target of rapamycin (mTOR) signaling. *J. Biol. Chem.* **277**, 23977–23980
55. Krause, U., Bertrand, L. & Hue, L. (2002) Control of p70 ribosomal protein S6 kinase and acetyl-CoA carboxylase by AMP-activated protein kinase and protein phosphatases in isolated hepatocytes. *Eur. J. Biochem.* **269**, 3751–3759
56. Nave, B.T., Ouwens, M., Withers, D.J., Alessi, D.R. & Shepherd, P.R. (1999) Mammalian target of rapamycin is a direct target for protein kinase B: identification of a convergence point for opposing effects of insulin and amino-acid deficiency on protein translation. *Biochem. J.* **344**, 427–431
57. Meisse, D., Van de Casteele, M., Beauloye, C., Hainault, I., Kefas, B.A., Rider, M.H., Foufelle, F. & Hue, L. (2002) Sustained activation of AMP-activated protein kinase induces C-jun N-terminal kinase activation and apoptosis in liver cells. *FEBS Lett.* **526**, 38–42
58. Motoshima, H., Golstein, B.J., Igata, M. & Araki, E. (2006) AMPK and cell proliferation-AMPK as a therapeutic target for atherosclerosis and cancer. *J. Physiol.* **574**, 63–71
59. Zhang, L., Li, J., Young, L.H. & Caplan, M.J. (2006) AMP-activated protein kinase regulates the assembly of epithelial tight junctions. *Proc. Natl. Acad. Sci. U.S.A.* **103**, 17272–17277
60. Blume, C., Benz, P.M., Walter, U., Ha, J., Kemp, B.E. & Renne, T. (2006) AMP-activated protein kinase impairs endothelial actin cytoskeleton assembly by phosphorylating vasodilator-stimulated phosphoprotein. *J. Biol. Chem.* **282**, 4601–4612

10

Hypoxia in the central nervous system

Joseph C. LaManna[1]

Department of Anatomy, Case Western Reserve University, School of Medicine, 10900 Euclid Avenue, Cleveland, OH 44106-4930 U.S.A.

Abstract

The brain, as a very high energy consumer, is completely reliant on molecular oxygen but because oxygen is dangerous due to toxicity [1], there are mechanisms which allow the brain to exist under low oxygen conditions when 'idling' but increase oxygen delivery when activated. This situation means that the brain can respond naturally to mild hypoxia with acute and chronic adaptive mechanisms. These mechanisms involve systemic and central metabolic and vascular processes that are mediated by hypoxia-inducible factor (HIF)-1. HIF-1-mediated cerebral angiogenesis is completed within 3 weeks of exposure onset and is reversible over the same time frame if normoxia is restored. Hypoxic acclimatizing responses may be significantly impaired with aging and metabolic or vascular disease.

Introduction

The mammalian brain depends completely on continuous delivery of oxygen to maintain its function. Because mammals populate varied environmental niches they are exposed to variable ambient oxygen levels. Therefore it should not be surprising that there are numerous intrinsic and extrinsic mechanisms that maintain a suitable level of oxygen availability to the neurons and other cells of the central nervous system. Hypoxia is any decrease in the availability of oxygen. When used alone, the term refers to a decrease in the PO_2 (partial pressure

[1]*To whom correspondence should be addressed (email joseph.lamanna@case.edu).*

of oxygen) in the ambient air from the mean sea level value of about 21 kPa (dry gas). In normal physiology, the PO_2 in the pulmonary vein is about 14 kPa due to the contribution of water vapour and carbon dioxide. Most of the oxygen is carried by haemoglobin and at these partial pressures, haemoglobin is fully saturated. Even if ambient oxygen decreases a little, haemoglobin remains fairly well-saturated and the tissue oxygen availability remains the same. When the arterial PO_2 falls below 12 kPa, which can occur through decreases in the F_iO_2 (fraction inspired oxygen), decreased barometric pressure such as increasing altitude, or through lung pathology (pulmonary hypoxia), then a condition of hypoxaemia (lower blood oxygen) occurs. A decrease in the oxygen availability to the tissue such as that which occurs in uncompensated hypoxaemia, anaemia (anaemic hypoxia, i.e. low haemoglobin and thus lower oxygen-carrying capacity), carbon monoxide (toxic hypoxia), or blood flow restriction (ischaemic hypoxia), all result in activation of local tissue response mechanisms.

For acute hypoxic exposures, if blood oxygen is below 12 kPa, but above 6 kPa, then the hypoxia is mild and can be compensated for by normal physiological processes, and usually does not lead to any tissue damage. Thus there are no problems for humans up to about 3000 m. In the range of 4–6 kPa, the hypoxia is moderate and is partially compensated, but there is good chance of permanent damage. Below 4 kPa, hypoxia is severe and loss of consciousness and neuronal degeneration ensues.

If mild hypoxia persists, long-term compensatory responses change the scenario. These acclimatizing adaptations may take days to weeks to become established, but then allow habitation at moderate hypoxia and brief periods of severe hypoxia with far less damage than before acclimatization. Thus, for example, a climber without supplementary oxygen might reach the peak at Mount Everest (elevation 8850 m), which is at an approximate pressure of one third of an atmosphere (about 34 kPa), with an estimated pulmonary alveolar PO_2 of 4.7 kPa, and an estimated arterial PO_2 of only 2.8 kPa.

Chronic exposure to the hypoxia of altitude over generations leads to selection for adaptations that increase suitability to the high-altitude environment. These adaptations allow human populations to inhabit areas above 4000 m in the Andes, in the Himalayas and in Ethiopia. Similarly, there are many other mammalian species that are adapted to even higher altitudes. Interestingly, the adaptations of these different species and human populations are varied.

As implied above, hypoxia itself is not necessarily a pathological condition. Indeed, the very reactivity that makes oxygen useful for energy-producing reactions bestows a potential for toxicity that must be accounted for by the organism. One of these protective strategies is to limit the exposure of the tissue to oxygen except for the amount needed to drive activity-induced metabolism. Thus the mammalian brain usually exists in a low tissue oxygen milieu. Local tissue mechanisms ensure that the oxygen environment is controlled, not maximized. These physiological considerations underlie the fMRI (functional

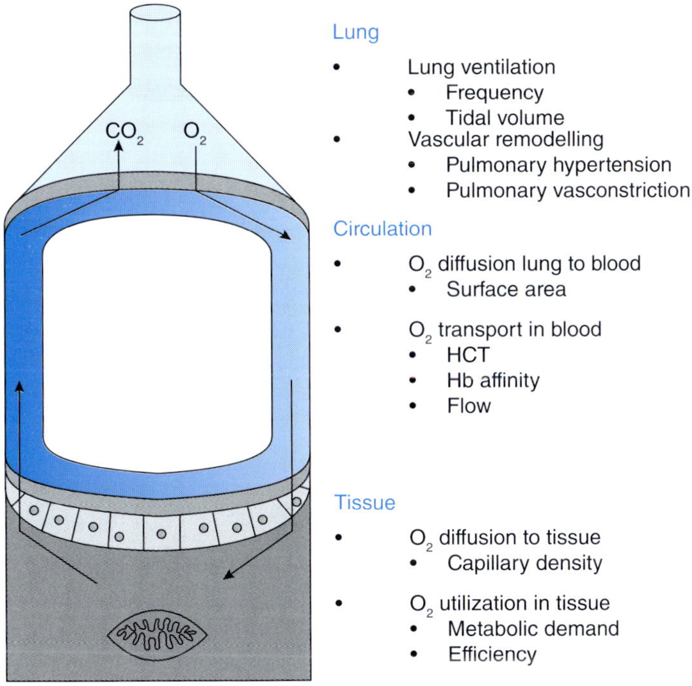

Figure 1. Scheme showing the major levels of acclimatization to prolonged mild hypoxia
Drawn by C. Tsipis, based on Weibel's diagram [2].

magnetic resonance imaging) phenomenon known as BOLD (blood oxygen level dependent) response to a focal activation task, that is proving to be a valuable tool for understanding brain function and has promise for pathological diagnosis and treatment monitoring.

This chapter concerns itself with the brain responses to mild hypoxia, that is, to physiological hypoxia. This is the range of hypoxia that can be compensated for with physiological mechanisms. More severe hypoxia leads to pathology and cell death due to failure of these compensation mechanisms and subsequent energy depletion.

Brain oxygen

The pathway for oxygen can be summarized by the diagram in Figure 1 [2]. There are three main levels at which adaptation to prolonged mild hypoxia can occur: the lung, the circulation and at the tissue level itself, in this case the brain parenchyma. The changes in the lung include an increase in diffusing capacity [3,4] and an increase in minute ventilation [5] (driven by carotid body oxygen-sensing cells through brain stem nuclei and phrenic output). Also there is pulmonary vasoconstriction and pulmonary hypertension (which lead to right heart hypertrophy), and vascular remodelling [6]. The circulatory contribution includes increased cardiac output, increased haemoglobin concentration and a right-shifted haemoglobin saturation curve. In the tissue, adaptations include increased capillary surface area, decreased energy demand and increased energy production efficiency.

The systemic mechanisms are fairly well understood. The lung, heart and kidney participate in the maintenance of oxygen delivery, defined as the product of blood oxygen content and blood flow, to the brain parenchyma through the level of the end arteriole.

Other systemic responses to prolonged mild hypoxia include decreased core temperature, bicarbonate ion excretion by the kidney to balance the blood alkalosis due to hyperventilation-induced carbon dioxide loss and body mass loss.

The effect of acute hypoxia in brain

The brain tissue parenchyma exists in a low oxygen environment. Quantitative measurements of brain tissue PO_2 by polarographic microelectrodes have shown that the distribution of oxygen tensions in brain is asymmetrical, shifted towards the lower concentrations, and best described with a log normal fit [7]. An example of the PO_2 distribution histogram is shown in Figure 2(A). Plotting these data as a cumulative occurrence function, as shown in Figure 2(B), makes it easier to observe that more than two-thirds of the values are below 2 kPa, and 90% are below the venous PO_2 partial pressure of 5 kPa. Decreases in inspired oxygen produce decreases in tissue PO_2. The response to, in this case eucapnic, hypoxia is shown in Figure 3. The capillary blood haemoglobin saturation falls, and with it the tissue PO_2. This decrease occurs despite an increased blood volume and faster capillary mean transit times as a result of vasodilation and increased cerebral blood flow [8]. The blood flow increase in mild hypoxia is at least partially mediated via oxygen-sensing neurons in the rostral ventrolateral medulla [9].

The failure of the acute central and systemic hypoxic responses to preserve tissue oxygen tension is a consequence of the limitations of oxygen diffusion from the capillaries to the tissue. Oxygen diffusion follows Fick's law which means that the flux at any distance from the red cell in the capillary is dependent on the concentration gradient for oxygen. Because tissue PO_2 is small, the oxygen flux rate is essentially dependent on the capillary PO_2. Thus, even though oxygen delivery in terms of molecules of oxygen per min at the end arteriole can be preserved, tissue oxygen tension must fall because arterial PO_2 has fallen.

Two responses to hypoxia are activated that resist this fall. With hypoxia there is an increase in ventilation rate. This is driven by the carotid body chemosensor pathway through brain stem nuclei. Ventilation rate increases acutely, then is maintained. An example is shown in Figure 4. Hyperventilation results in a decreased PCO_2 (partial pressure of carbon dioxide) which allows the arterial oxygen tension to rise equivalently. This results in an increase of about

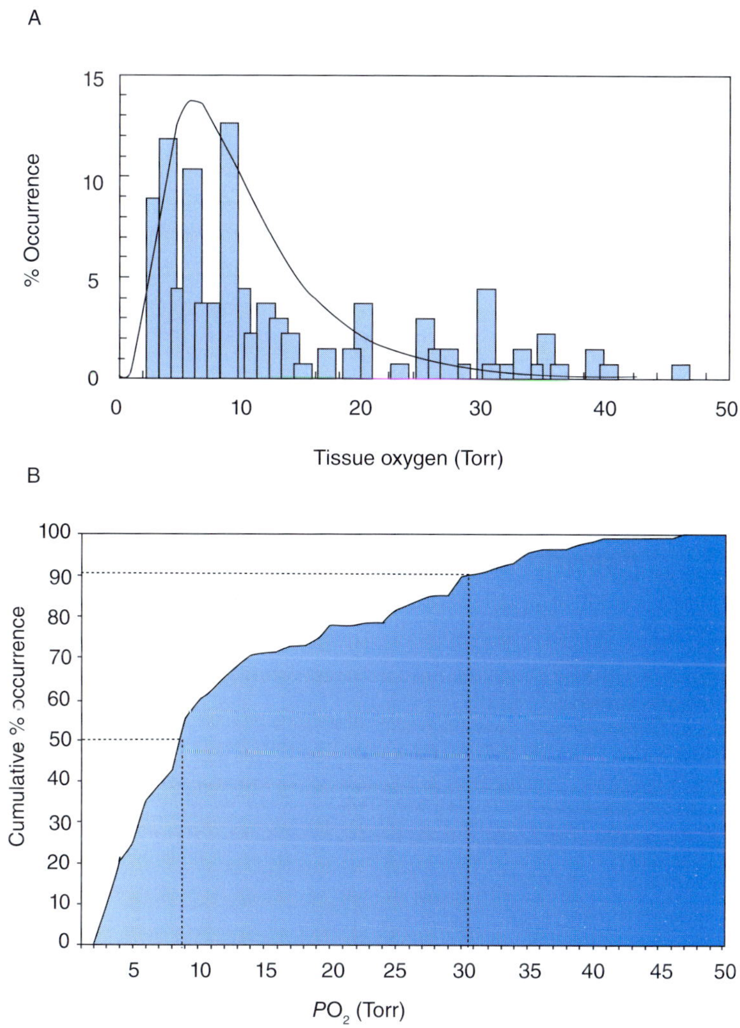

Figure 2. Cerebral cortical tissue oxygen distribution
(**A**) Histogram distribution of tissue oxygen tension measurements made from the cerebral cortices of 13 anaesthetized rats. The data have been replotted from [7]. A log normal distribution is indicated by the line. (**B**) The same data plotted as a cumulative distribution showing the PO_2 for the 50 and 90 percentiles.

10% in the oxygen-driving force at the arterial end of the capillary. The carotid body responds to PO_2 [5], or at least, the minute ventilation continues to be inversely proportional to the PO_2.

The other adaptation has a slower onset occurring only after weeks of exposure [10]. If mild hypoxic exposure is maintained, angiogenesis is initiated which results in an increased capillary branching and density which results in decreased intercapillary distances and thus decreased diffusion distances, restoring tissue oxygen tensions to baseline.

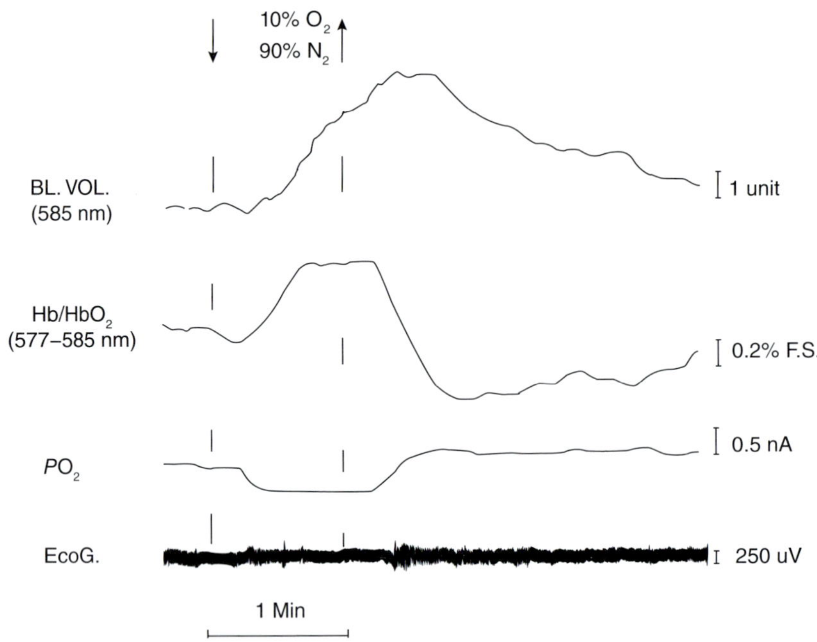

Figure 3. Hypoxia-induced acute changes in blood volume, haemoglobin saturation and tissue oxygen tension

The response to forced eucapnic ventilation with 10% oxygen on the oxygen tension, measured by an oxygen-sensitive platinum polarographic microelectrode; blood volume and tissue capillary haemoglobin saturation, measured by *in vivo* reflectance spectrophotometry; and the surface electrocorticogram, measured by the Ag-AgCl reference on the polarographic electrode, in anaesthetized rat cortex. This figure is from previously unpublished work collected during the studies reported in [43].

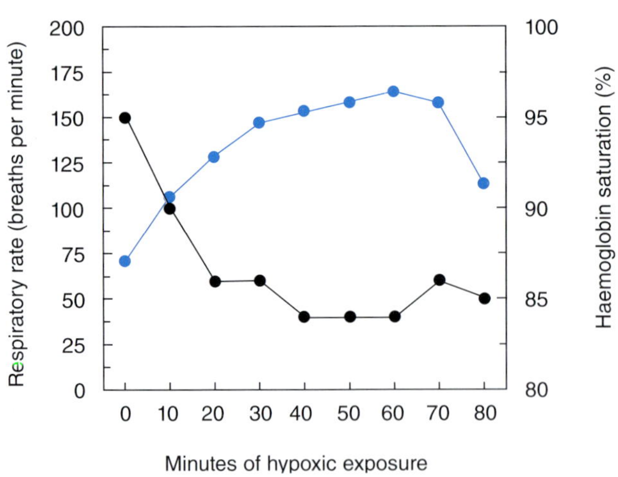

Figure 4. Blood oxygen saturation (black) and ventilation rate (blue) measured non-invasively in a restrained rat placed in a 10% oxygen environment

Chronic exposure

A primary player in hypoxic angiogenesis is hypoxia-inducible factor (HIF)-1 [11]. HIF-1 is a heterodimeric transcription factor with a constituitive HIF-1β, identical with ARNT (aryl hydrocarbon receptor nuclear translocator), and an inducible component HIF-1α. The levels of HIF-1α increase with hypoxia due to the inhibition of the oxygen requiring PHD (prolyl hydroxylase) in the breakdown pathway, and can thus serve as an oxygen sensor [12]. HIF-1 is a transcription factor that activates over 40 known genes that have an HRE (hypoxic-response element) in their promoter region. VEGF (vascular endothelial growth factor) is one such molecule that is upregulated by HIF-1 and initiates capillary angiogenesis.

Angiogenesis begins within the first week of hypoxic exposure and continues for at least another week. By 3 weeks the remodelling is complete. The result is an increase in capillary density. An example is shown in Figure 5.

HIF-1α remains responsive to tissue PO_2, it does not adapt. The HIF-1 signal is thus maintained until the tissue PO_2 is restored by angiogenesis.

Competent angiogenesis also requires participation by a second, non-HIF-1-dependent mechanism that involves up-regulation of COX-2 (cyclo-oxygenase-2) and subsequent up-regulation of angiopoietin-2 via PGE_2 (prostaglandin E_2) receptor activation [13,14]. The mechanism of up-regulation of COX-2 in hypoxia is not yet known.

Systemically, HIF-1 is responsible for hypoxic erythropoiesis. HIF-1 accumulation due to hypoxia in the kidney results in up-regulation of EPO (erythropoietin) that results in an increase, over days to a week, in packed red cell volume, that contributes to the oxygen-carrying capacity of the arterial blood [15]. In rats exposed to an equivalent altitude of 5500 m, where the PO_2 is about half that at sea level, the packed red cell volume is increased to 70% or more compared with less than 50% at sea level. As the oxygen-carrying capacity increases, the cerebral blood flow falls back to pre-exposure levels. This

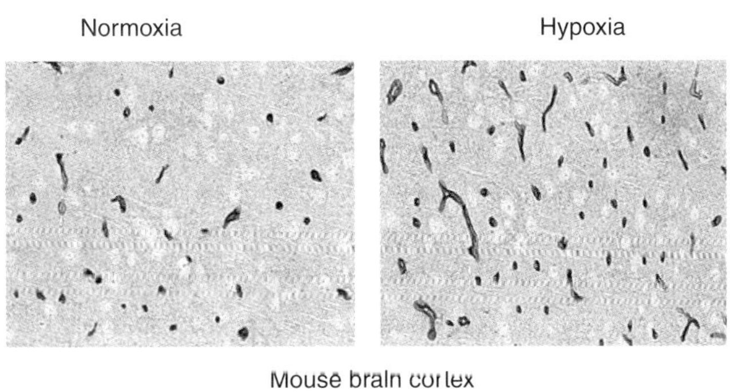

Figure 5. A section from the cortex of a mouse exposed for 3 weeks to hypobaric hypoxia compared with a littermate kept in normoxic conditions
The hypoxic cortex shows a higher density of capillaries (stained immunohistologically for GLUT1, which is a specific marker for brain capillaries, at the blood–brain barrier).

occurs before brain tissue oxygen tension has been restored. Thus the sensor for this mechanism must be responsive to arterial oxygen content, not PO_2 in the blood or tissue [16].

Almost all of the enzymes involved in glycolysis have an HRE and are up-regulated in prolonged mild hypoxia [17]. The cerebral metabolic rate for glucose is elevated in both acute [8,18] and chronic hypoxia [19]. The increase of about 10–20% in CMRglu (cerebral metabolic rate for glucose) would contribute a relatively small amount to energy needs of the cells. Glycolysis produces almost twenty times less ATP per glucose molecule than oxidative phosphorylation, so a 20% increase would be negligible. Nevertheless, the increase in glycolysis appears to balance the tissue pH which would otherwise tend to become alkalotic due to the decrease in arterial PCO_2 [20]. Systemic alkalosis is balanced by secretion of bicarbonate in the kidney, but in the central nervous system alkalosis is offset by increased glycolytic ATP production. In oxidative phosphorylation, a proton is consumed when ATP is synthesized. In glycolysis, there is no net production or consumption of protons. When ATP is hydrolysed in energy-requiring reactions, a proton is produced. Thus utilization of ATP produced from oxidative phosphorylation is pH neutral and consumption of ATP from glycolysis is acid-producing [21]. Since glycolysis provides the substrate for oxidative phosphorylation, there are always some protons being produced, but an increase in glycolysis without a corresponding increase in oxidative phosphorylation will increase acid production, and this is demonstrated by an increase in brain lactate concentration.

Not only does the glycolytic rate increase, but the transport of glucose at the blood–brain barrier is also increased [22]. The GLUT1 (glucose transporter 1), responsible for carrier-mediated facilitated diffusion at the blood–brain barrier is also up-regulated by HIF-1 [11]. In the rat, the increase in glucose transporter and the increase in capillary density result in a 3-fold increase in glucose flux rate capacity in hypoxic-adapted rats. The large disparity in the increased transport capacity compared with the increased CMRglu can be explained by the net reduction by half of the arterial glucose delivery. In species such as the rat (but not humans) that lack GLUT1 in circulating red blood cells, the source of glucose available for transport is limited to the glucose in the plasma. After hypoxic adaptation, the rate of whole blood flow through the tissue re-normalizes, but this means that the plasma flow rate is half the pre-exposure rate. The increased transport capacity compensates for the decreased plasma flow rate, supports the increased resting CMRglu, and allows for transient increases in energy demand due to focal neuronal activation.

Other changes in the tissue include a decrease in the volume density of neuronal mitochondria [23] and cytochrome oxidase activity [24,25] that probably reflect an overall decrease in resting $CMRO_2$ (cerebral metabolic rate for oxygen) of about 15%, although this has not been measured directly. The tendency for hypoxia to decrease brain activity and metabolism may be related to the idea of central respiratory depression [26]. An overall decrease in body temperature [27] may reflect decreased metabolism as an acclimatization strategy that includes a drop in body mass.

Intrinsic brain tissue oxygen sensors

The HIF-1 pathway clearly plays a role as an oxygen sensor in the brain. HIF-1 accumulates in hypoxia due to inhibition of PHD activity which requires oxygen. On the other hand, hypoxia is not the only agent resulting in HIF-1 accumulation. For example, non-hypoxic HIF-1 accumulation can occur due to growth factors such as IGF-1 (insulin-like growth factor 1), iron chelation, cobalt chloride and alteration of substrate availability. Likewise, overproduction or stimulation of PHD will prevent HIF-1 accumulation even in moderate hypoxia.

Other sensor candidates would be oxygen-dependent enzymes, such as tryptophan and tyrosine hydroxylases [28], with affinity constants in the range of physiological tissue oxygen tensions, i.e. less than 2 kPa. But monoamine metabolism shows only a transient decrease in activity that reverses after 36 h.

There are also iron-containing molecules that might act as sensors because they bind oxygen in the physiological range. For example, neuroglobin, a haem protein, is up-regulated by sustained hypoxia and may play a protective role [29,30]. The effector mechanisms of these iron-containing proteins have not as yet been investigated [31].

Species differences

Hypoxia-induced angiogenesis has not been studied in many different species. It is important to establish the fundamental principles of hypoxic mammalian brain physiology and distinguish mechanisms that are species specific. Hypoxic HIF-1 accumulation appears to be universal in all cells that have a competent metabolic apparatus, so a great deal of the response should be common. On the other hand, there are significant differences among vertebrates and even in special mammalian cases, such as diving mammals, with respect to strategies in dealing with hypoxia [32,33].

Environmental niche is certain to have a major influence on hypoxic response. Diving mammals are just one example of this. Burrowing animals which spend more time in high carbon dioxide environments, ungulates which have different dietary patterns, and animals that have existed at altitude for countless generations each will exhibit variations [34,35]. High-altitude animals have evolved a left-shifted haemoglobin saturation curve which facilitates haemoglobin loading in the lung as an adaptive trait, the opposite of low-land animals which respond to prolonged hypoxia with a right-shift to facilitate oxygen unloading in the tissues [36]. Studies on the capillary density in these species should be of interest.

Another important consideration, often overlooked, is the relationship of body mass to the pertinent physiological variables. For example, in addition to well-known scaling of brain blood flow and brain metabolism, the oxygen dissociation curve apparently also scales with body mass [37], the smaller animals have a more right-shifted *p*50 (oxygen partial pressure producing 50% saturation).

Rats have similar physiological responses to hypoxia as humans for a number of variables [38], but the rat haemoglobin *p*50 is right-shifted compared with humans. Also unknown is the effect of GLUT1 which is quite prominent in human erythrocytes but absent in small mammals such as rats and mice.

Effects of re-oxygenation

As interesting as the hypoxia-induced angiogenesis is, what happens on return to normoxia is even more intriguing. Angiopoietin-2 is again found to be elevated, but HIF-1 and its downstream growth factors are not [13]. Under these conditions of re-oxygenation, a rarefaction of capillaries occurs through a process of physiological apoptosis. Thus it appears that the COX-2-mediated pathway is activated during re-modelling to either increase or decrease capillary density, depending on the presence or absence of other factors [39]. The process of angiogenesis in the brain requires a remodelling signal because the capillaries are usually so tightly connected mechanically. In the case of hypoxia-induced angiogenesis, this involves angiopoietin-2, from COX-2 and PGE_2, to prepare the vessels for remodelling. In the case of re-oxygenation, the remodelling signal is activated, but the growth factors from the HIF-1 system are missing. This has the effect of initiating removal of the microvessels through physiological apoptosis.

The remodelling of the cerebral capillary vasculature in response to changes in tissue oxygen implies a dynamic process of continual matching of local energy demand and energy supply in the tissue, with a response time of days to weeks. Appreciation of this process is still rudimentary. It is likely that these mechanisms participate in the modification of capillary density and neuronal metabolism seen in rats raised in complex sensory environments or in the motor training that accompanies prolonged exercise [40,41]. But the sensors, signals and effectors remain mostly unknown.

Effects of age

Finally, it should be noted that most of the research on brain hypoxia involves healthy young subjects. There is, however, known age sensitivities to hypoxia that could reveal profound implications for dynamic angiogenesis and be most useful for understanding the acclimatization process. At least in the case of older rats, HIF-1 fails to accumulate in mild hypoxia [42]. This does not occur through failure of synthesis since HIF-1 mRNA is functional, and HIF-1 accumulation does occur in response to iron chelators or cobalt chloride treatment. It appears that there is a primary deficit in the oxygen-sensing function, most likely through overproduction of PHD. These older rats should have a much more difficult time acclimatizing to hypoxia, learning motor tasks or surviving metabolic stresses such as cerebral ischaemia.

Conclusions

Mild, prolonged hypoxia evokes systemic and central nervous system mechanisms that result in successful acclimatization. In mammals, the central nervous system response includes increased glucose metabolism, decreased oxidative capacity and microvascular remodelling by capillary angiogenesis which results in decreased diffusion distances for oxygen from erythrocytes to mitochondria. The changes induced by hypoxia are reversible upon return to a normal oxygen environment. The robust response indicates that the brain vasculature is not fixed but dynamically adjusted to the supply-and-demand conditions of the functional tissue units. The transcription factor, HIF-1, plays a major role in orchestrating the metabolic and vascular responses to hypoxia, but clearly does not act alone. Failure or inadequacy of this complex acclimatization process may underlie many of the pathological outcomes of metabolic and vascular diseases, especially in the aging population. It will be important to determine the specific characteristics of the hypoxic acclimatization process in humans and to investigate strategies for interventions that could promote health and recovery.

Summary

- *The brain normally exists in a low oxygen environment.*
- *Brain tissue oxygen tension is spatially and temporally heterogeneous.*
- *There are systemic and central mechanisms that help compensate for acute brain hypoxia; (i) an increase in lung ventilation mediated by carotid body sensors that respond to arterial* PO_2*; (ii) an increase in cerebral blood flow mediated by brain stem sensors that respond to arterial oxygen content; and (iii) an increased brain parenchymal capillary density mediated by local sensors that respond to tissue* PO_2*.*
- *Angiogenesis occurs when hypoxia persists for a few weeks. These adaptive changes are reversible. These changes are impaired in aging.*

This author's research has been continually supported by the National Institutes of Health, currently under R01-NS38632. Constantinos Tsipis, Xiaoyan Sun, Bernadette Erokwu and Kui Xu helped with the previously unpublished experiments, diagrams and figures.

References

1. Acker, H. (2005) The oxygen sensing signal cascade under the influence of reactive oxygen species. *Philos. Trans. R. Soc. London Ser. B.* **360**, 2201–2210

2. Weibel, E.R. (1984) *The Pathway for Oxygen*, Harvard University Press, Cambridge
3. Bartlett, Jr, D. & Remmers, J.E. (1971) Effects of high altitude exposure on the lungs of young rats. *Respir. Physiol.* **13**, 116–125
4. Burri, P.H. & Weibel, E.R. (1971) Morphometric estimation of pulmonary diffusion capacity. II. Effect of PO_2 on the growing lung, adaption of the growing rat lung to hypoxia and hyperoxia. *Respir. Physiol.* **11**, 247–264
5. Wilson, D.F., Roy, A. & Lahiri, S. (2005) Immediate and long-term responses of the carotid body to high altitude. *High Alt. Med. Biol.* **6**, 97–111
6. Howell, K., Preston, R.J. & McLoughlin, P. (2003) Chronic hypoxia causes angiogenesis in addition to remodelling in the adult rat pulmonary circulation. *J. Physiol.* **547**, 133–145
7. Sick, T.J., Lutz, P.L., LaManna, J.C. & Rosenthal, M. (1982) Comparative brain oxygenation and mitochondrial redox activity in turtles and rats. *J. Appl. Physiol.* **53**, 1354–1359
8. Beck, T. & Krieglstein, J. (1987) Cerebral circulation, metabolism, and blood–brain barrier in rats in hypocapnic hypoxia. *Am. J. Physiol.* **252**, H504–H512
9. Reis, D.J., Golanov, E.V., Ruggerio, D.A. & Sun, M.-K. (1994) Sympatho-excitatory neurons of the rostral ventrolateral medulla are oxygen sensors and essential elements in the tonic and reflex control of the systemic and cerebral circulation. *J. Hypertens.* **12** (Suppl 10), S159–S180
10. LaManna, J.C., Vendel, L.M. & Farrell, R.M. (1992) Brain adaptation to chronic hypobaric hypoxia in rats. *J. Appl. Physiol.* **72**, 2238–2243
11. Chávez, J.C., Agani, F., Pichiule, P. & LaManna, J.C. (2000) Expression of hypoxic inducible factor 1α in the brain of rats during chronic hypoxia. *J. Appl. Physiol.* **89**, 1937–1942
12. Sharp, F.R. & Bernaudin, M. (2004) HIF1 and oxygen sensing in the brain. *Nat. Rev. Neurosci.* **5**, 437–448
13. Pichiule, P. & LaManna, J.C. (2002) Angiopoietin-2 and rat brain capillary remodeling during adaptation and de-adaptation to prolonged mild hypoxia. *J. Appl. Physiol.* **93**, 1131–1139
14. Pichiule, P., Chavez, J.C. & LaManna, J.C. (2004) Hypoxic regulation of angiopoietin-2 expression in endothelial cells. *J. Biol. Chem.* **279**, 12171–12180
15. Semenza, G.L. & Wang, G.L. (1992) A nuclear factor induced by hypoxia via de novo protein synthesis binds to the human erythropoietin gene enhancer at a site required for transcriptional activation. *Mol. Cell. Biol.* **12**, 5447–5454
16. Xu, K. & LaManna, J.C. (2006) Chronic hypoxia and the cerebral circulation. *J. Appl. Physiol.* **100**, 725–730
17. LaManna, J.C., Chavez, J.C. & Pichiule, P. (2007) Genetics and gene expression: glycolysis. In *Handbook of Neurochemistry & Molecular Neurobiology, 3rd Edition*, (Gibson, G. & Dienel, G. eds), pp. 771–788, Springer, New York
18. Pulsinelli, W.A. & Duffy, T.E. (1979) Local cerebral glucose metabolism during controlled hypoxemia in rats. *Science* **204**, 626–629
19. Harik, S.I., Lust, W.D., Jones, S.C., Lauro, K.L., Pundik, S. & LaManna, J.C. (1995) Brain glucose metabolism in hypobaric hypoxia. *J. Appl. Physiol.* **79**, 136–140
20. Lauro, K.L. & LaManna, J.C. (1997) Adequacy of cerebral vascular remodeling following three weeks of hypobaric hypoxia. Examined by an integrated composite analytical model. *Adv. Exp. Med. Biol.* **411**, 369–376
21. Dennis, S.C., Gevers, W. & Opie, L.H. (1991) Protons in ischemia: where do they come from; where do they go to? *J. Mol. Cell. Cardiol.* **23**, 1077–1086
22. Harik, S.I., Behmand, R.A. & LaManna, J.C. (1994) Hypoxia increases glucose transport at blood–brain barrier in rats. *J. Appl. Physiol.* **77**, 896–901
23. Stewart, P.A., Isaacs, H., LaManna, J.C. & Harik, S.I. (1997) Ultrastructural concomitants of hypoxia-induced angiogenesis. *Acta Neuropathol.* **93**, 579–584
24. LaManna, J.C., Kutina-Nelson, K.L., Hritz, M.A., Huang, Z. & Wong-Riley, M.T.T. (1996) Decreased rat brain cytochrome oxidase activity after prolonged hypoxia. *Brain Res.* **720**, 1–6
25. Chávez, J.C., Pichiule, P., Boero, J. & Arregui, A. (1995) Reduced mitochondrial respiration in mouse cerebral cortex during chronic hypoxia. *Neurosci. Lett.* **193**, 169–172

26. Neubauer, J.A., Melton, J.E. & Edelman, N.H. (1990) Modulation of respiration during brain hypoxia. *J. Appl. Physiol.* **68**, 441–451
27. Wood, S.C. (1991) Interactions between hypoxia and hypothermia. *Ann. Rev. Physiol.* **53**, 71–85
28. Davis, J.N. (1975) Adaptation of brain monoamine synthesis to hypoxia in the rat. *J. Appl. Physiol.* **39**, 215–220
29. Li, R.C., Lee, S.K., Pouranfar, F., Brittian, K.R., Clair, H.B., Row, B.W., Wang, Y. & Gozal, D. (2006) Hypoxia differentially regulates the expression of neuroglobin and cytoglobin in rat brain. *Brain Res.* **1096**, 173–179
30. Sun, Y., Jin, K., Peel, A., Mao, X.O., Xie, L. & Greenberg, D.A. (2003) Neuroglobin protects the brain from experimental stroke *in vivo*. *Proc. Natl. Acad. Sci. U.S.A.* **100**, 3497–3500
31. Brunori, M. & Vallone, B. (2006) A globin for the brain. *FASEB J.* **20**, 2192–2197
32. Hochachka, P.W. & Lutz, P.L. (2001) Mechanism, origin, and evolution of anoxia tolerance in animals. *Comp. Biochem. Physiol. B. Biochem. Mol. Biol.* **130**, 435–459
33. Hochachka, P.W. (1986) Defense strategies against hypoxia and hypothermia. *Science* **231**, 234–241
34. Llanos, A.J., Riquelme, R.A., Sanhueza, E.M., Hanson, M.A., Blanco, C.E., Parer, J.T., Herrera, E.A., Pulgar, V.M., Reyes, R.V., Cabello, G. & Giussani, D.A. (2003) The fetal llama versus the fetal sheep: different strategies to withstand hypoxia. *High Alt. Med. Biol.* **4**, 193–202
35. Gilbert, R.D., Pearce, W.J. & Longo, L.D. (2003) Fetal cardiac and cerebrovascular acclimatization responses to high altitude, long-term hypoxia. *High Alt. Med. Biol.* **4**, 203–213
36. Schmidt-Nielsen, K. (1983) *Animal Physiology: Adaptation and Environment*, Cambridge University Press, New York
37. Schmidt-Nielsen, K. & Larimer, J.L. (1958) Oxygen dissociation curves of mammalian blood in relation to body size. *Am. J. Physiol.* **195**, 424–428
38. Musch, T.I., Dempsey, J.A., Smith, C.A., Mitchell, G.S. & Bateman, N.T. (1983) Metabolic acids and [H^+] regulation in brain tissue during acclimatization to chronic hypoxia. *J. Appl. Physiol.* **55**, 1486–1495
39. Holash, J., Maisonpierre, P.C., Compton, D., Boland, P., Alexander, C.R., Zagzag, D., Yancopoulos, G.D. & Wiegand, S.J. (1999) Vessel cooption, regression, and growth in tumors mediated by angiopoietins and VEGF. *Science* **284**, 1994–1998
40. Swain, R.A., Harris, A.B., Wiener, E.C., Dutka, M.V., Morris, H.D., Theien, B.E., Konda, S., Engberg, K., Lauterbur, P.C. & Greenough, W.T. (2003) Prolonged exercise induces angiogenesis and increases cerebral blood volume in primary motor cortex of the rat. *Neurosci.* **117**, 1037–1046
41. Black, J.E., Sirevaag, A.M. & Greenough, W.T. (1987) Complex experience promotes capillary formation in young rat visual cortex. *Neurosci. Lett.* **83**, 351–355
42. LaManna, J.C., Chavez, J.C. & Pichiule, P. (2004) Structural and functional adaptation to hypoxia in the rat brain. *J. Exp. Biol.* **207**, 3163–3169
43. Kreisman, N.R., Sick, T.J., LaManna, J.C. & Rosenthal, M. (1981) Local tissue oxygen tension - cytochrome a,a_3 redox relationships in rat cerebral cortex in vivo. *Brain Res.* **218**, 161–174

11

Hypoxia and Alzheimer's disease

Chris Peers*[1], Hugh A. Pearson† and John P. Boyle*

**Faculty of Medicine, University of Leeds, Leeds LS2 9JT, U.K. and †Faculty of Biological Sciences, University of Leeds, Leeds LS2 9JT, U.K.*

Abstract

Numerous cardiorespiratory disorders result in persistent systemic hypoxia, or at worst (as a consequence of stroke) deprive the brain of oxygen completely for a period of time. Patients suffering from such conditions are much more susceptible to the development of dementias such as AD (Alzheimer's disease). Until recently, the cellular and molecular basis for the predisposition to AD by systemic hypoxia has been completely unknown. However, emerging evidence suggests that pathological cellular remodelling caused by chronic hypoxia shows striking similarities to those observed in the central nervous system as a consequence of AD. Furthermore, prolonged hypoxia can induce formation of Aβs (amyloid β peptides), the primary neurotoxic elements of AD, which accumulate over years to form the extracellular plaques that are the hallmark feature of the disease. Hypoxia can lead to paradoxical increases in mitochondrial ROS (reactive oxygen species) generation upstream of Aβ formation. The downstream consequences of prolonged hypoxia include remodelling of functional expression of voltage-gated calcium channels and disturbance of intracellular calcium homoeostasis via disrupted calcium buffering and inhibition of calcium extrusion mechanisms. These effects can be mimicked by application of exogenous

[1]*To whom correspondence should be addressed (email c.s.peers@leeds.ac.uk).*

Aβ and, crucially, appear to depend on Aβ formation. Current knowledge supports the concept that prevention of the deleterious effects of hypoxia may prove beneficial in slowing or preventing the onset of AD.

Introduction

Dementias, of which AD (Alzheimer's disease) is by far the most common, are characterized by a progressive decline in memory, which becomes increasingly associated with confusion over simple, everyday tasks. As time passes, symptoms worsen and problems with memory become more obvious, as does confusion and general cognitive impairment. In later stages, mood swings, depression and anger are apparent and general mental deterioration can progress ultimately to death. Dementias are most commonly associated with old age; indeed, senile dementia specifically refers to patients over 65, and the risk of developing dementia doubles every 5 years after this age. The current global statistics surrounding dementia are alarming: worldwide there are an estimated 24–25 million people suffering from dementia. Each year, 4.6 million new cases arise (one every 7 s), which will lead to an estimated doubling of sufferers every 20 years [1]. This is a consequence of the global aging of the general population since there is no evidence that the rate at which AD presents in 65-year-olds and over is increasing. The preceding facts are in themselves shocking, but the impact on society is worsened by socioeconomic factors: the cost of caring for AD patients is greater than the *combined* cost of caring for cancer, stroke and heart disease patients. AD is clearly a major, worldwide current and future concern.

A brief scan through the clinical literature will reveal numerous reports which collectively suggest that diseases associated with a lowering of systemic oxygen levels predispose individuals to AD (reviewed briefly in [2]). The most striking, but by no means exclusive, data are from patients who have recovered from a stroke. The likelihood of such patients subsequently developing dementias is several-fold higher than it is for age-matched subjects who have not faced such an insult. One of the clearest accounts is the longitudinal study by Desmond et al. [3] who found the incidence of dementias increased approximately 6-fold in stroke patients (Figure 1). Importantly, they concluded that the risk of dementia amongst stroke patients was particularly high if associated with complications resulting in subsequent cerebral hypoxia or ischaemia. Despite this study, and others resulting in similar conclusions, no cellular or molecular basis for the association of hypoxia/ischaemia with AD had been forthcoming until approximately 5 years ago. This chapter will review the data which are accumulating to account for this association, but first commences with a brief account of the biochemical basis for AD, and current models for neuronal death which underlie this disease.

Cellular production of amyloid peptides

The hallmark feature of AD is the presence of extracellular plaques in the brain, along with intracellular neurofibrillary tangles (composed of

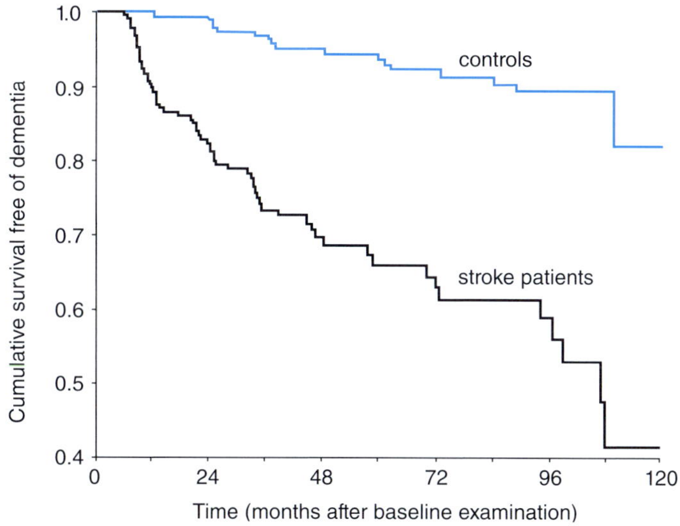

Figure 1. Kaplan–Meier analysis showing cumulative proportion of subjects (either controls or stroke patients, as indicated) surviving free of dementia, stratified by stroke status during follow-up of up to 120 months
Reproduced from [3], Desmond, D.W., Moroney, J.T., Sano, M. & Stern, Y. (2002) Incidence of dementia after ischemic stroke: results of a longitudinal study. Stroke **33**, 2254–2260 with permission from Lippincott, Williams and Wilkins.

hyperphosphorylated tau protein, see [4] for recent review). Although tangles are clearly of great importance in the disease, most research has focused on the major constituent of the extracellular plaques, Aβ (amyloid β peptide), since most evidence suggests this peptide is responsible for neuronal death in AD. Aβ is a cleavage product formed by the sequential actions of two enzymes on APP (amyloid precursor protein). APP is a ubiquitously expressed, membrane-spanning protein which can undergo alternative cleavage pathways as illustrated in Figure 2. When APP is cleaved by α-secretase, this occurs within the Aβ sequence, precluding Aβ formation and instead yielding sAPPα, which is believed to be neuroprotective [5]. The identity (or identities) of the α-secretase remain to be fully clarified, but most evidence suggests it/they belong to the ADAM (**A** **D**isintegrin **A**nd **M**etalloprotease) family of proteases [6]. Subsequent cleavage of the α-secretase product yields additional fragments which may be biologically active [2]. The alternative amyloidogenic processing of APP involves the sequential activity of two enzymes, β- and γ-secretase (Figure 2). β-secretase, also known as BACE-1 (β-amyloid cleavage enzyme), Asp2 or memapsin2, is a single protein of the aspartyl protease family. By contrast, γ-secretase is a protein complex, consisting of presenilin (−1 or −2, where the cleavage site resides), nicastrin, Aph-1 and Pen-2 [7]. Both β- and γ-secretase present attractive therapeutic targets in AD, since both are required to liberate Aβ.

It is important to note that Aβ formation is a physiological process, and there is accumulating evidence that the peptide serves important physiological roles [2]. Levels of Aβ are maintained by its clearance and degradation, the latter

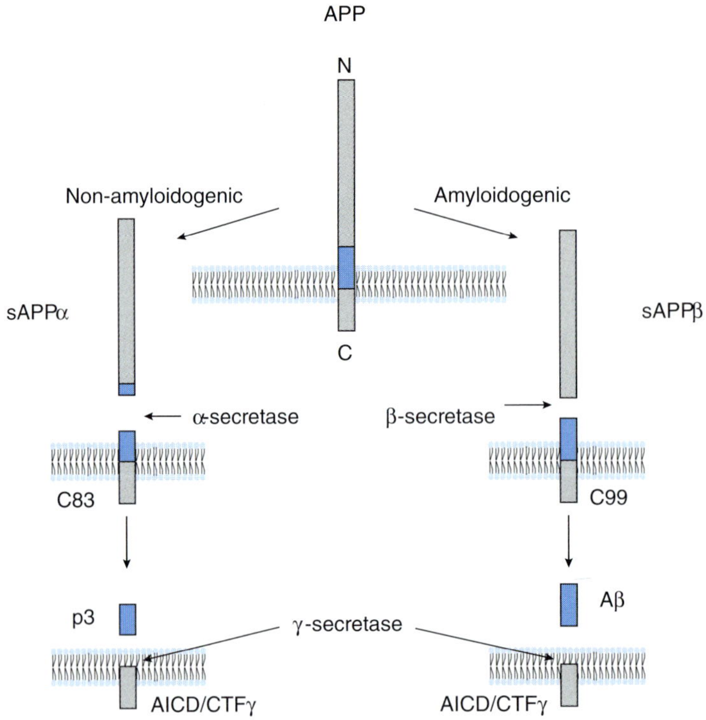

Figure 2. Schematic illustrating the proteolytic processing of APP
Proteolytic processing of APP occurs via either the non-amyloidogenic processing pathway (left; to generate the neuroprotective sAAPα) or the amyloidogenic pathway (right; to generate Aβ). In both pathways, other biologically active fragments are also generated. AICD, APP intracellular domain; CTF, C-terminal fragment.

being a function of peptidases such as insulin-degrading enzyme, endothelin-converting enzyme and neprilysin [8]. An imbalance between the production and the removal of Aβ is believed, according to the amyloid cascade hypothesis, to account for the development of neuronal dysfunction and, ultimately, death in AD. The stages of disease progression from initiation of amyloid burden to final deposition in plaques, and the consequences for neurones and other central nervous system cells during this period, currently form the basis of research interests of a large community of scientists whose common aim is to understand, and so prevent, this growing disease.

Cellular dysfunction and death in AD

The development of extracellular amyloid plaques and intraneuronal neurofibrillary tangles of hyperphosphorylated tau are well documented features of AD progression. So too is the loss of synaptic integrity and neuronal death in limbic and associated cortical regions of the brain. Less well understood, however, is the cellular and molecular basis for neuronal death. Clearly, evidence points towards Aβ causing apoptotic cell death, but the route of destruction

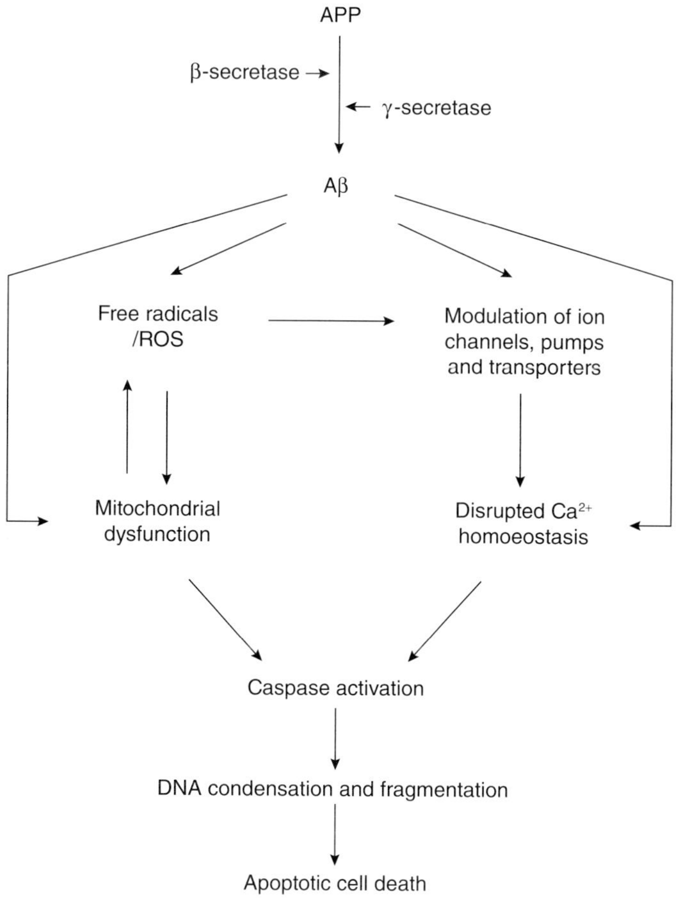

Figure 3. Schematic showing simplified possible routes to apoptotic cell death caused by Aβ

Central to the scheme are the generation of ROS and disruption of calcium homoeostasis. Evidence suggests Aβ can affect by both direct and indirect means. Note that dysfunctional mitochondria are both a source and a target of ROS. Both primary mechanisms, ROS production and disruption of calcium homoeostasis, display crosstalk at several points, and converge at the level of caspase activation to promote apoptosis.

remains to be fully evaluated [9,10]. Nevertheless, a vast literature of studies addressing this issue has arisen over the past 15 years or so, and from such studies, some clear themes arise pointing to major pathways involved in cell death in AD. These are summarized in Figure 3. In brief, there are two fundamental mechanisms of primary importance in Aβ-induced neuronal death, one involving free radical/ROS (reactive oxygen species) generation, and the other centering on calcium dyshomoeostasis [9,11,12]. The whole process is complicated by the fact that these two pathways are inextricably linked at many levels. Thus, for example, increased ROS production can disrupt the functional expression of various types of ion channels, pumps and transporters, any of which can contribute to calcium homoeostasis. Disruption of calcium homoeostasis (which can even occur due to the action of presenilin, independently of its ability to generate Aβ; [13]) can lead to mitochondrial dysfunction which can lead in turn to increased ROS formation. The two pathways ultimately lead to DNA condensation and fragmentation via caspase activation, but this is by no means the only point at which these enzymes act in this pathway [10]. Further complexity is added to the system by additional factors including the involvement of NF-κB (nuclear factor κB; which can be protective in neurones, but destructive via oxidative damage when activated in glia [14]) and a protective role for elevated calcium levels via PKB (protein kinase B)/Akt activation [15]. Different aspects of these pathways are championed by various laboratories, but in reality it seems we are some way yet from determining which mechanism is of primary importance, or whether both suggested pathways actually represent one complex mechanism for cell death in AD.

Hypoxic promotion of Aβ formation

Until recently, despite the clinical associations of hypoxia/ischaemia with AD, the cellular basis for such an association was unknown. However, the primary mechanisms of cellular dysfunction associated with Aβ (increased ROS production and disruption of calcium signalling; Figure 3) share numerous commonalities with the effects of hypoxia, prompting an examination into mechanisms by which hypoxia and AD may be linked at the cellular and molecular level. Much of our own group's efforts have been aimed at understanding this linkage, and the fundamental observation which we have made is that exposing cells in tissue culture to periods of hypoxia (1–2.5% oxygen for 6–48 h) increases the production of amyloid peptides. We have demonstrated this in a variety of cell types, including PC12 cells, HEK (human embryonic kidney)-293 cells and the human neuroblastoma, SH-SY5Y [16–18], as well as in primary cultures of cortical astrocytes [19,20] and, importantly, central neurones [21]. Thus this common feature seems to occur regardless of cell type. However, the mechanism by which this occurs does appear cell-specific. For example, in SH-SY5Y cells, hypoxia promotes suppression of ADAM10 (a candidate α-secretase) expression and decreases sAPPα secretion [22], but in cortical astrocytes hypoxia is pro-amyloidogenic via increased expression of presenilin-1, a key component of the γ-secretase complex [20]. Our knowledge of how hypoxia promotes amyloidogenic APP processing is still in its infancy, but the consequences of such effects have been studied in more depth, shedding more light on the mechanistic links between the damaging effects of hypoxia and the neurodegeneration of AD. Of these, we shall focus on two key aspects of pathological remodelling of cell function; (i) calcium channel expression and (ii) calcium signalling pathways.

Remodelling of calcium channel functional expression

Early studies demonstrated that chronic (24 h) hypoxia, in addition to promoting Aβ formation (see above), also potentiated depolarization-evoked, calcium-dependent neurotransmitter release, due to augmentation of calcium influx [16,23]. In PC12 cells, this was confirmed using patch-clamp recordings: hypoxia augmented the whole-cell calcium current, an effect entirely due to selective, functional up-regulation of L-type (Cav1) calcium channels, despite the presence in these cells of other voltage-activated calcium channels [24]. Importantly, this effect has also been demonstrated in primary cultures of central neurones [21]. Furthermore, hypoxic increases in whole cell calcium channel currents were also observed in HEK-293 cells stably expressing the α_{1C} subunit of the human L-type calcium channel, Cav1.2 [17]. Use of a recombinant expression system permitted detailed examination of the underlying mechanisms, which are summarized in Figure 4. Immediately, the reproduction of the effects of hypoxia in a recombinant expression system makes

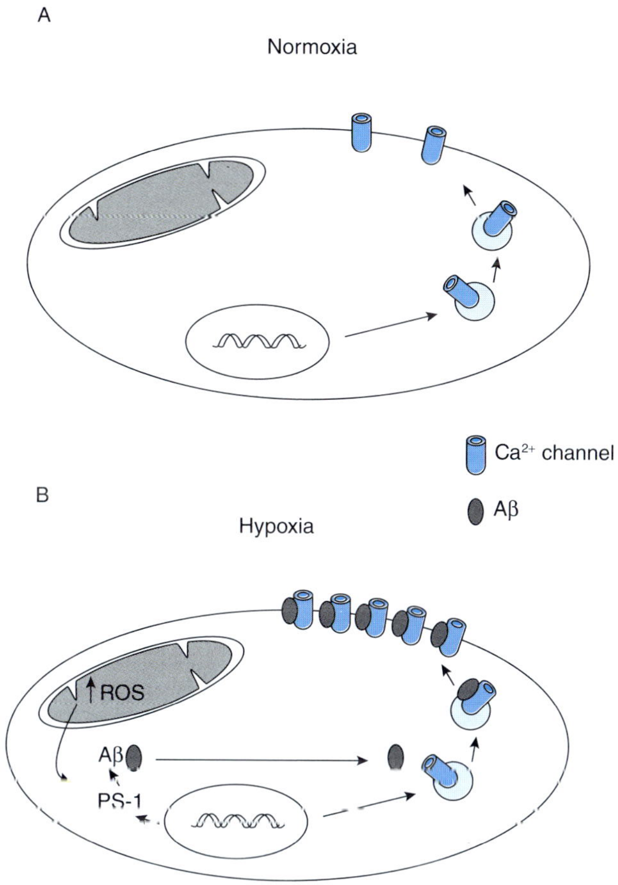

Figure 4. Hypoxic modulation of calcium channel expression
In normoxia (**A**) calcium channel synthesis, trafficking and membrane insertion are depicted. This process is disrupted in hypoxia (**B**) due to increased ROS production and Aβ formation. Evidence points to a possible direct interaction of Aβ with the calcium channel protein, disrupting trafficking via an unknown mechanism which results in more active channels present in the plasma membrane. PS-1, presenilin-1.

the possibility that hypoxia increases calcium channel transcription highly unlikely. Instead, we observed an increase in post-transcriptional trafficking of L-type channels such that at any given time more were present and active in the plasma membrane, even though total cell channel content was not markedly altered. Crucially, these effects of hypoxia were inhibited by selective inhibitors of either β- or γ-secretase, indicating an absolute requirement for Aβ in this process [17]. Indeed, the altered trafficking may have resulted from a close or even direct interaction of Aβ with the calcium channel subunit, since the two proteins co-localized and co-immunoprecipitated [17].

The above-described process provides an attractive mechanism by which hypoxia could contribute to cell death in AD via disruption of calcium homoeostasis, as outlined in Figure 2. There is also evidence for the involvement of ROS production in this phenomenon, since antioxidants prevented hypoxic up-regulation of calcium channels in the same recombinant expression system [18]. Perhaps surprisingly, ROS involvement in this mechanism appears to be upstream of Aβ formation and the ROS appeared to be derived from the mitochondrial respiratory chain, since effects were inhibited by rotenone (an electron chain transport inhibitor) and by functional mitochondrial depletion induced by sustained exposure to ethidium bromide (cells derived this way are termed ρ^0 cells). Furthermore, exposure of ρ^0 cells to oxidative stress, or to Aβ itself, reproduced the effects of hypoxia to increase calcium channel up-regulation [18]. Increased ROS production from mitochondria has long been associated with aging, and appears to be intimately involved with the progression of AD [25]. Furthermore, there is a growing body of evidence to suggest ROS production from mitochondria paradoxically increases during hypoxia [26]. Collectively, these data provide an attractive scheme to account for a mechanism by which hypoxia could contribute to both calcium- and ROS-mediated neuronal damage or death in AD, via disruption of calcium channel functional expression.

Hypoxic disruption of calcium signalling

Although voltage-gated calcium entry is of paramount importance in neurones, other mechanisms for regulating calcium influx and, indeed, intracellular calcium buffering are equally important. Indeed, in non-excitable cells such as astrocytes, voltage-gated calcium entry has little or no physiological role, and calcium signalling proceeds by entirely different mechanisms. Agonists binding to phospholipase C-coupled receptors generate IP_3 (inositol trisphosphate) which can mobilize calcium from the ER (endoplasmic reticulum) which in turn activates capacitative- (or store-operated) calcium entry. These means of calcium signalling are fundamental to numerous functions in the central nervous system and other tissues [27,28]. Exposure of various cell types to chronic hypoxia caused apparent augmentation of agonist-evoked mobilization of calcium from ER stores and additionally modulates capacitative calcium entry [29–31]. This has been studied in depth in primary cultures of astrocytes, cells which rely heavily for many functions on intra- and inter-cellular calcium wave propagation [32]. These effects are summarized in Figure 5. In essence, prolonged hypoxia augments agonist-evoked rises of [calcium]$_i$ not by increasing the ER calcium content, but by inhibiting clearance of calcium from the cytosol, thereby causing its accumulation. Inhibition arose in part from suppressed expression of the plasmalemmal Na^+/Ca^{2+} exchanger [33], but also from calcium-loading of mitochondria. This latter effect, in turn, appeared to be due to hyperpolarization of the mitochondrial membrane potential upon which

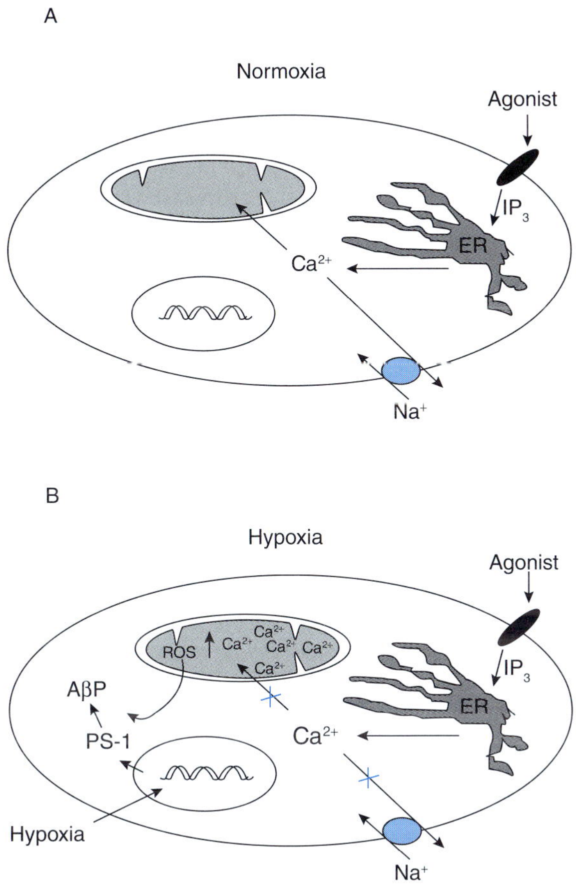

Figure 5. Hypoxic modulation of calcium signalling

Schematic indicating agonist-mediated alterations in [calcium]$_i$ in normoxia (**A**) and in chronic hypoxia (**B**). In both cases, agonist application liberates calcium from the ER, which is normally buffered by mechanisms including extrusion on the Na^+/Ca^{2+} exchanger and uptake into mitochondria (**A**). Following chronic hypoxia (**B**), these two buffering mechanisms are prevented, leading to larger rises of [calcium]$_i$ when cells are exposed to the agonist. The mechanism of Na^+/Ca^{2+} exchanger inhibition is unknown, but the ability of mitochondria to buffer rises of $[Ca^{2+}]_i$ is due to the fact that chronic hypoxia has already caused them to become overloaded with calcium through hyperpolarization [30]. These damaging changes are associated with increased ROS levels derived from mitochondria, and increased Aβ production, presumably as a result of increased levels of the pro-amyloidogenic presenilin-1 (PS-1).

calcium uptake relies [31]. Again, these effects were seen to be associated with, indeed reliant upon, increased Aβ formation and generation of ROS [20]. Thus, as with remodelling of calcium channel functional expression, perturbations in calcium homoeostasis and signalling caused by hypoxia involves pro-amyloidogenic processes.

Conclusions

It is clear from numerous clinical studies that conditions which disrupt oxygen delivery to the central nervous system (of which stroke is the most severe and well-documented example) promote the onset of dementias in general, and AD in particular. At the cellular level, restriction of oxygen supply promotes formation of ROS from mitochondria and disrupts the functional expression of numerous proteins associated with calcium homoeostasis, including ion channels and transporters. Strikingly, ROS production and calcium dyshomoeostasis are two recognized pathways leading to neuronal death in AD and, indeed, hypoxia also promotes formation of Aβ. These cellular studies provide potential mechanisms to account for the association of hypoxic/ischaemic episodes and increased incidence of AD. In so doing, they provide clues as to targets which might prove useful in ameliorating the long-term consequences of such episodes.

Summary

- *AD is a neurodegenerative disease in which neuronal death occurs via complex pathways involving generation of ROS and disturbance of intracellular calcium homoeostasis.*
- *Hypoxic or ischaemic events, such as stroke, dramatically increase the likelihood of developing AD.*
- *Hypoxia promotes the formation of Aβ, the primary neurotoxic element of AD.*
- *Many of the adverse effects of prolonged hypoxia on cellular processes crucial to intracellular calcium homoeostasis are mediated by increased ROS production and require Aβ formation.*
- *Current data suggest that oxygen availability may be a major determinant of Aβ formation and that mechanisms contributing to hypoxic/ischaemic cellular damage and death are similar to those of AD.*

The authors' contributions to the work described herein have been supported by The Medical Research Council, The Wellcome Trust, The Alzheimer's Research Trust and the Alzheimer's Society. We apologise to colleagues in the field whose work we could not cite due to limitations of space.

References

1. Ferri, C.P., Prince, M., Brayne, C., Brodaty, H., Fratiglioni, L., Ganguli, M., Hall, K., Hasegawa, K., Hendrie, H., Huang, Y. et al. (2005) Global prevalence of dementia: a Delphi consensus study. *Lancet* **366**, 2112–2117
2. Pearson, H.A. & Peers, C. (2006) Physiological roles for amyloid β peptides. *J. Physiol.* **575**, 5–10
3. Desmond, D.W., Moroney, J.T., Sano, M. & Stern, Y. (2002) Incidence of dementia after ischemic stroke: results of a longitudinal study. *Stroke* **33**, 2254–2260
4. Blennow, K., de Leon, M.J. & Zetterberg, H. (2006) Alzheimer's disease. *Lancet* **368**, 387–403
5. Hooper, N.M. (2005) Roles of proteolysis and lipid rafts in the processing of the amyloid precursor protein and prion protein. *Biochem. Soc. Trans.* **33**, 335–338
6. Vardy, E.R., Catto, A.J. & Hooper, N.M. (2005) Proteolytic mechanisms in amyloid-β metabolism: therapeutic implications for Alzheimer's disease. *Trends Mol. Med.* **11**, 464–472
7. Dominguez, D.I., Hartmann, D. & De, S.B. (2004) BACE1 and presenilin: two unusual aspartyl proteases involved in Alzheimer's disease. *Neurodegener. Dis.* **1**, 168–174
8. Saido, T.C. & Iwata, N. (2006) Metabolism of amyloid β peptide and pathogenesis of Alzheimer's disease. Towards presymptomatic diagnosis, prevention and therapy. *Neurosci. Res.* **54**, 235–253
9. Mattson, M.P., LaFerla, F.M., Chan, S.L., Leissring, M.A., Shepel, P.N. & Geiger, J.D. (2000) Calcium signaling in the ER: its role in neuronal plasticity and neurodegenerative disorders. *Trends Neurosci.* **23**, 222–229
10. Mattson, M.P. (2000) Apoptosis in neurodegenerative disorders. *Nat. Rev. Mol. Cell Biol.* **1**, 120–129
11. LaFerla, F.M. (2002) Calcium dyshomeostasis and intracellular signalling in Alzheimer's disease. *Nat. Rev. Neurosci.* **3**, 862–872
12. Reddy, P.H. (2006) Amyloid precursor protein-mediated free radicals and oxidative damage: implications for the development and progression of Alzheimer's disease. *J. Neurochem.* **96**, 1–13
13. Tu, H., Nelson, O., Bezprozvanny, A., Wang, Z., Lee, S.F., Hao, Y.H., Serneels, L., De, S.B., Yu, G. & Bezprozvanny, I. (2006) Presenilins form ER Ca^{2+} leak channels, a function disrupted by familial Alzheimer's disease-linked mutations. *Cell* **126**, 981–993
14. Mattson, M.P. (2005) NF-κB in the survival and plasticity of neurons. *Neurochem. Res.* **30**, 883–893
15. Yano, S., Tokumitsu, H. & Soderling, T.R. (1998) Calcium promotes cell survival through CaM-K kinase activation of the protein-kinase-B pathway. *Nature* **396**, 584–587
16. Taylor, S.C., Batten, T.F. & Peers, C. (1999) Hypoxic enhancement of quantal catecholamine secretion. Evidence for the involvement of amyloid β-peptides. *J. Biol. Chem.* **274**, 31217–31222
17. Scragg, J.L., Fearon, I.M., Boyle, J.P., Ball, S.G., Varadi, G. & Peers, C. (2005) Alzheimer's amyloid peptides mediate hypoxic up-regulation of L-type Ca^{2+} channels. *FASEB J.* **19**, 150–152
18. Brown, S.T., Scragg, J.L., Boyle, J.P., Hudasek, K., Peers, C. & Fearon, I.M.,(2005) Hypoxic augmentation of Ca^{2+} channel currents requires a functional electron transport chain. *J. Biol. Chem.* **280**, 21706–21712
19. Peers, C., Smith, I.F., Boyle, J.P. & Pearson, H.A. (2004) Remodelling of Ca^{2+} homeostasis in type I cortical astrocytes by hypoxia: evidence for association with Alzheimer's disease. *Biol. Chem.* **385**, 285–289
20. Smith, I.F., Boyle, J.P., Green, K.N., Pearson, H.A. & Peers, C. (2004) Hypoxic remodeling of Ca^{2+} mobilization in type I cortical astrocytes: involvement of ROS and pro-amyloidogenic APP processing. *J. Neurochem.* **88**, 869–877
21. Webster, N.J., Ramsden, M., Boyle, J.P., Pearson, H. & Peers, C. (2006) Amyloid peptides mediate hypoxic increase of L-type Ca^{2+} channels in central neurones. *Neurobiol. Aging* **27**, 439–445
22. Webster, N.J., Green, K.N., Peers, C. & Vaughan, P.F. (2002) Altered processing of amyloid precursor protein in the human neuroblastoma SH-SY5Y by chronic hypoxia. *J. Neurochem.* **83**, 1262–1271
23. Webster, N.J., Vaughan, P.F. & Peers, C. (2001) Hypoxic enhancement of evoked noradrenaline release from the human neuroblastoma SH-SY5Y. *Brain Res. Mol. Brain Res.* **89**, 50–57
24. Green, K.N. & Peers, C. (2001) Amyloid β peptides mediate hypoxic augmentation of Ca^{2+} channels. *J. Neurochem.* **77**, 953–956

25. Lin, M.T. & Beal, M.F. (2006) Mitochondrial dysfunction and oxidative stress in neurodegenerative diseases. *Nature* **443**, 787–795
26. Bell, E.L., Emerling, B.M. & Chandel, N.S. (2005) Mitochondrial regulation of oxygen sensing. *Mitochondrion* **5**, 322–332
27. Verkhratsky, A. (2005) Physiology and pathophysiology of the calcium store in the endoplasmic reticulum of neurons. *Physiol. Rev.* **85**, 201–279
28. Putney, Jr, J.W. (2005) Capacitative calcium entry: sensing the calcium stores. *J. Cell Biol.* **169**, 381–382
29. Smith, I.F., Boyle, J.P., Vaughan, P.F., Pearson, H.A. & Peers, C. (2001) Effects of chronic hypoxia on Ca^{2+} stores and capacitative Ca^{2+} entry in human neuroblastoma (SH-SY5Y) cells. *J. Neurochem.* **79**, 877–884
30. Smith, I.F., Plant, L.D., Boyle, J.P., Skinner, R.A., Pearson, H.A. & Peers, C. (2003) Chronic hypoxia potentiates capacitative Ca^{2+} entry in type I cortical astrocytes. *J. Neurochem.* **85**, 1109–1116
31. Smith, I.F., Boyle, J.P., Plant, L.D., Pearson, H.A. & Peers, C. (2003) Hypoxic remodeling of Ca^{2+} stores in type I cortical astrocytes. *J. Biol. Chem.* **278**, 4875–4881
32. Scemes, E. & Giaume, C. (2006) Astrocyte calcium waves: what they are and what they do. *Glia* **54**, 716–725
33. Atkinson, L., Boyle, J.P., Pearson, H.A. & Peers, C. (2006) Chronic hypoxia inhibits Na^{+}/Ca^{2+} exchanger expression in cortical astrocytes. *Neuroreport* **17**, 649–652

12

Hypoxia in cancer cell metabolism and pH regulation

M. Christiane Brahimi-Horn[1] and Jacques Pouysségur

Institute of Signaling, Developmental Biology and Cancer Research, CNRS UMR 6543, University of Nice, Centre A. Lacassagne, 33 Avenue Valombrose, 06189 Nice, France

Abstract

At a molecular level, hypoxia induces the stabilization and activation of the α-subunit of an α/β heterodimeric transcription factor, appropriately termed HIF (hypoxia-inducible factor). Hypoxia is encountered, in particular, in tumour tissues, as a result of an insufficient and defective vasculature present in a highly proliferative tumour mass. In this context the active HIF heterodimer binds to and induces a panel of genes that lead to modification in a vast range of cellular functions that allow cancer cells to not only survive but to continue to proliferate and metastasize. Therefore HIF plays a key role in tumorigenesis, tumour development and metastasis, and its expression in solid tumours is associated with a poor patient outcome. Among the many genes induced by HIF are genes responsible for glucose transport and glucose metabolism. The products of these genes allow cells to adapt to cycles of hypoxic stress by maintaining a level of ATP sufficient for survival and proliferation. Whereas normal cells metabolize glucose through a cytoplasmic- and mitochondrial-dependent pathway, cancer cells preferentially use a cytoplasmic, glycolytic pathway that leads to an increased acid load due, in part, to the high level of production of lactic acid.

[1]*To whom correspondence should be addressed (email brahimi@unice.fr).*

This metabolic predilection of cancer cells is primarily dependent directly on the HIF activity but also indirectly through changes in the activity of tumour suppressors and oncogenes. A better understanding of HIF-dependent metabolism and pH regulation in cancer cells should lead to further development of diagnostic tools and novel therapeutics that will bring benefit to cancer patients.

Introduction

The term hypoxia is used to describe a 'low' PO_2 (partial pressure of oxygen), whereas normoxia refers to a 'normal' PO_2. However, in tissues what is hypoxic to one tissue may be normoxic to another and vice versa, since different tissues have different normoxic levels of oxygen, which vary according to the degree of vascularization [1] and degree of oxygen consumption. In an *in vitro* cellular situation HIF (hypoxia-inducible factor)-α becomes progressively more stable as the PO_2 decreases from 40 mmHg (around 5% oxygen) to anoxia. The tumour PO_2 (1–10 mmHg) lies well within this range [2] as does that for certain normal tissues such as in the brain, retina and skin. The stability of HIF-α in hypoxia is explained by the inhibition of its destruction by the proteasomal machinery. In the presence of oxygen, HIF-α is highly unstable due to successive post-translational hydroxylation and ubiquitination that earmarks it for proteasomal capture. Stable HIF-α translocates to the nucleus where it dimerizes with HIF-β and binds to site-specific DNA sequences on target genes. Over 70 genes, involved in a vast array of cellular functions, are induced or repressed in a HIF-dependent manner [3–5]. In tumours, the induction of genes such as the *vegf* (vascular endothelial growth factor) and *ang-2* (angiopoietin-2) reflects an attempt by tumour cells to re-establish vascular delivery of nutrients and oxygen through the promotion of the growth of new blood vessels from an existing network, a process referred to as angiogenesis; thereby continued cell proliferation is assured. The recognition of this angiogenic, adaptive response and its inhibition has lead to the development of effective drug treatments for several types of cancer [6]. Targeting of other HIF-dependent gene products and/or HIF-induced modifications to the tumour micro-environment should also lead to the development of efficacious drug treatments, but a full understanding of the hypoxic signalling cascade in tumours is essential [7].

The HIF family

HIF is composed of an α- and β-subunit (Figure 1). These proteins belong to a large family of proteins termed bHLH (basic-helix-loop-helix) proteins containing a PAS (Per-Arnt-Sim) domain that plays diverse functions, including in the response to hypoxia [8]. Three α-subunit isoforms, coded by different genes, together with a number of splice variants have been described. Research efforts have focused primarily on HIF-1α and more recently on HIF-2α whereas HIF-3α, the lesser-studied isoform, exists as six different splice variants [9], one of which plays a dominant-negative regulator role

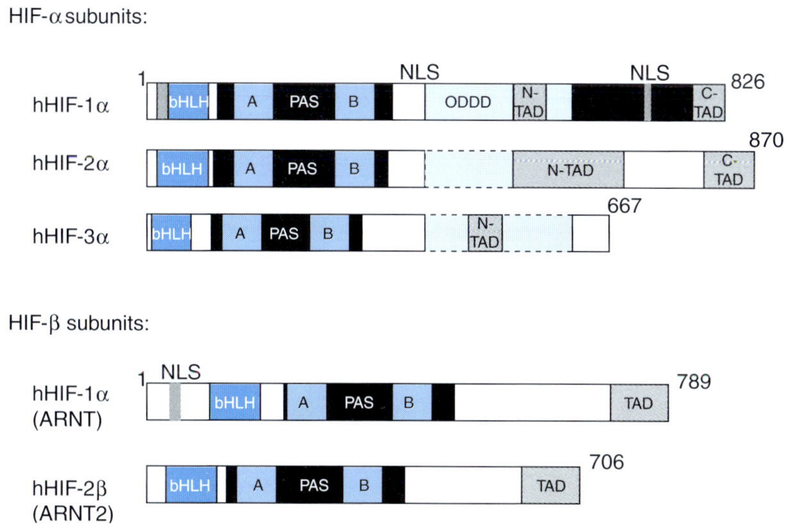

Figure 1. Domain structure of the HIF subunits

HIF consists of an α- and β-subunit both of which belong to the bHLH family of proteins containing PAS domains. Three human isoforms of the α-subunit, numbered 1–3, and two of the β-subunit, also termed ARNT (aryl hydrocarbon receptor nuclear translocator), are involved in the hypoxic response of cells. The labile α-subunits contain an ODDD (oxygen-dependent degradation domain) that determines proteasomal degradation. TADs reside in the C-terminal region and a NLS (nuclear localization signal) in the N-terminal region have been identified on certain isoforms.

[10]. Comparison of the sequence identity between HIF-1α and -2α shows 48% overall identity, with the N-terminal region showing the highest similarity. HIF-3α also shows considerable similarity in the bHLH-PAS region compared with the other isoforms. Both HIF-1α and -2α are expressed in a broad range of tissues but do show a degree of tissue specificity, in particular in leucocytes, endothelial cells and smooth muscle cells [8]. HIF-1β is also expressed in a broad range of tissues, whereas the lesser-studied HIF-2β is highly expressed in developing mouse brain [11]. The β-subunit can associate with other α-class bHLH-PAS proteins involved in a number of cellular functions, including detoxification, tracheal development and neurogenesis [8]. This multi-association, leading to multi-functions, suggests that the β-subunit can be limiting under certain circumstances. The α-subunits are highly labile in the presence of oxygen while the β-subunits are not regulated by oxygen. The inherent instability of HIF-α results from successive post-translational hydroxylation and ubiquitination leading to proteasomal degradation.

Regulation of the stability of HIF-α through post-translational modification

The major mechanism for regulation of HIF-α stability involves coupled hydroxylation and ubiquitination, but other forms of post-translational

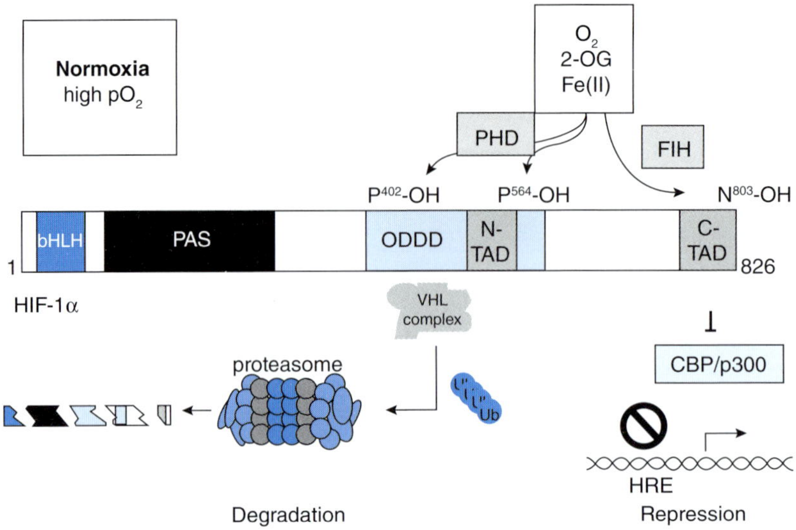

Figure 2. Oxygen sensor regulation of the stability and activity of HIF-1α
Under normoxic (high PO_2) conditions the oxygen sensor PHD proteins and FIH are active and hydroxylate HIF-1α on proline and asparagine residues respectively. These dioxygenases require oxygen, 2-OG and Fe^{2+} for activity. Consequently HIF-1α is labile and inactive through the interaction with the VHL–ubiquitin E3-containing complex that ubiquitinates (Ub) HIF-1α, earmarking it for degradation by the proteasome and abrogating interaction with the co-activator p300/CBP.

modification, including acetylation and sumoylation, have been reported to influence either stability or activity [12,13] however, this awaits further confirmation. The mechanism of post-translational regulation of stability is similar for the different α-subunits and is controlled by oxygen-requiring dioxygenases [14]. PHD (prolyl hydroxylase domain) dioxygenases hydroxylate proline residues in the oxygen-dependent degradation domain of HIF-α (Pro^{402} and Pro^{564} in human HIF-1α). This modification signals recruitment to HIF-α of an E3 ubiquitin ligase complex containing the VHL (von Hippel-Lindau tumour-suppressor) protein (Figure 2). VHL, together with its elongin B, elongin C, Cul 2 and Rbx-1 ubiquitinates HIF-α earmarking it for proteasomal destruction. Mutations in the *vhl* gene leading to loss-of-function are associated with RCC (renal cell carcinoma) and VHL disease, a familial syndrome [15]. HIF-α is therefore stable in these cancers and HIF target genes are activated, as reflected by the very high vascularization of these tumours. These characteristics also provide a strong link between HIF and angiogenesis and tumorigenesis.

HIF-dependent gene induction

Post-translational hydroxylation of HIF-α regulates not only stability but also transcriptional activity. A dioxygenase termed FIH (factor inhibiting HIF) controls the transcriptional activity of the α/β heterodimer through hydroxylation of an asparagine residue in the C-terminal region of HIF-1α

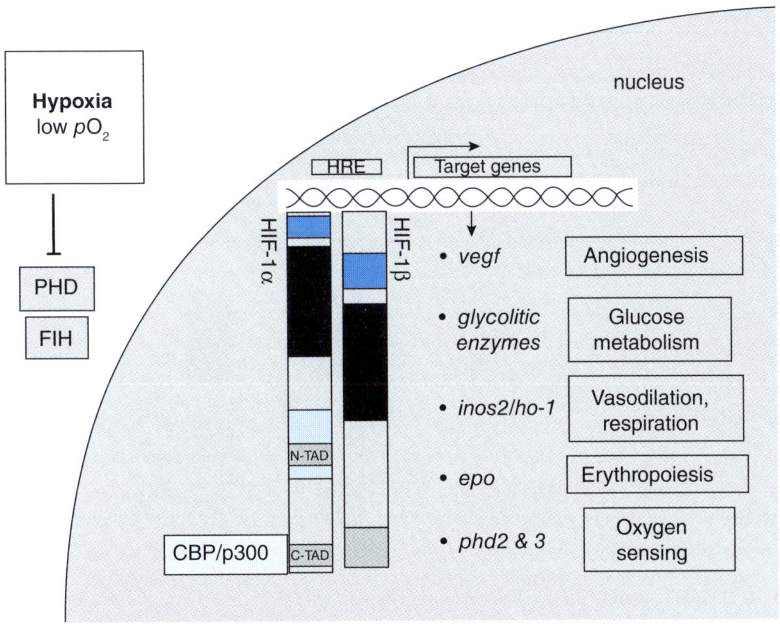

Figure 3. Hypoxic induction of HIF transcriptional activity
Under hypoxic (low PO_2) conditions the oxygen sensor PHD proteins and FIH are inactive due to diminished oxygen concentrations. The α-subunit of HIF is stable and translocates to the nucleus where it heterodimerizes with the β-subunit and binds to DNA at an HRE in enhancer or promoter regions of genes. The co-activator p300/CBP binds to the CAD (C-TAD) of HIF-α and activates the transcription of numerous genes including *vegf*, glycolytic enzymes, *ho-1* (haemoxygenase-1), *i-nos2* (inducible nitric oxide synthase), *epo* (erythropoietin) and *phd2* and *phd3* that are implicated in angiogenesis, metabolism, vasodilation/respiration, erythropoiesis and the hypoxic response respectively.

and -2α [16]. Hydroxylation by FIH abrogates interaction with a histone acetyltransferase co-activator p300/CBP (cAMP-response-element-binding protein-binding protein) (Figure 2). FIH is predominantly located in the cytoplasm of cells and its expression is not regulated by the oxygen concentration. Thus under low PO_2 HIF-α is stable and translocates to the nucleus where it associates with HIF-β through their respective N-terminal bHLH-PAS domains. The heterodimer then targets genes containing site-specific HREs (hypoxia-response elements), present on a large number of genes, and transcriptionally activates or represses their expression [3–5] (Figure 3).

The transcriptional activity of HIF is determined by two TADs (transcriptional activation domains), the NAD (N-terminal TAD) and CAD (C-terminal TAD), located in the C-terminal region of HIF-1α and -2α; where only the CAD is negatively regulated by FIH. Investigation into the significance of the existence of two TADs revealed that different genes are more or less dependent on FIH inhibition, thus on one or the other, or both, TAD [17] (Figure 4). *In vitro* assays indicate that PHD and FIH are active under different ranges of PO_2, whith PHD requiring a higher PO_2 for activity. Since an oxygen gradient

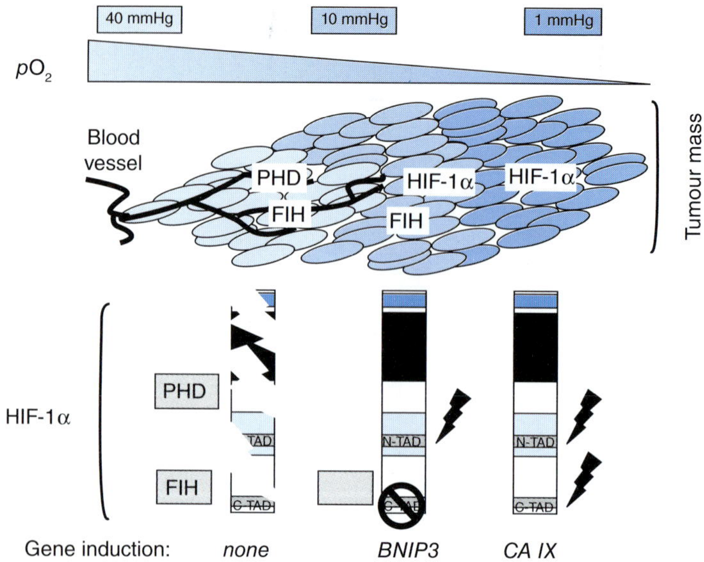

Figure 4. Bifunctional transcriptional activation of HIF

A gradient of PO_2 from normoxic (40 mmHg), to mild hypoxic (10 mmHg) and finally to severe hypoxic (1 mmHg) conditions occurs as tumour cells proliferate away from blood vessels. *In vitro* data suggest that in mild and severe hypoxia the PHD proteins are inhibited while FIH is inhibited only under severe hypoxia. Thus in normoxia, HIF-1α is labile and progressively becomes stable and active as the PO_2 diminishes. Since HIF-1α possesses two TADs, where the CAD (C-TAD) activity is regulated by post-translational hydroxylation by FIH, it can be hypothesized that in mild hypoxia FIH maintains a lock on the CAD (C-TAD) activity but that FIH-independent transcription via the NAD (N-TAD) can occur. In severe hypoxia both the NAD (N-TAD) and CAD (C-TAD) are activated. Experimental evidence suggests that different genes may be more or less dependent on one or both of the two TADs where, for example, *BNIP3* is NAD (N-TAD) and *CA IX* NAD (N-TAD)- and CAD (C-TAD)-dependent.

exists in tumours, a situation where HIF-α is stable but inactive in its CAD, due to FIH hydroxylation, is therefore hypothetically valid as illustrated by differential dependence of different genes on FIH [17]. Thus this double-headed property in transcriptional control represents a gene-selective mechanism for fine-tuning of the hypoxic response in tumours.

HIF isoform selectivity represents another gene-selective mechanism. This is illustrated by the finding that the HIF-dependent gene *ca9* (carbonic anhydrase 9) is predominantly induced by HIF-1 whereas HIF *phd3* (prolyl hydroxylase 3) is induced by HIF-1 and -2 [18].

Regulation of the metabolic balance in cancer cells by HIF

Histological examination of solid tumours generally shows a central necrotic region surrounded by a ring of viable cells that are poorly vascularized. The progression towards necrosis in a tumour mass is associated with a decreasing gradient in the level of oxygen and nutrients, and an increasing gradient toward a lower pH. To compensate for a low oxygen and nutrient supply,

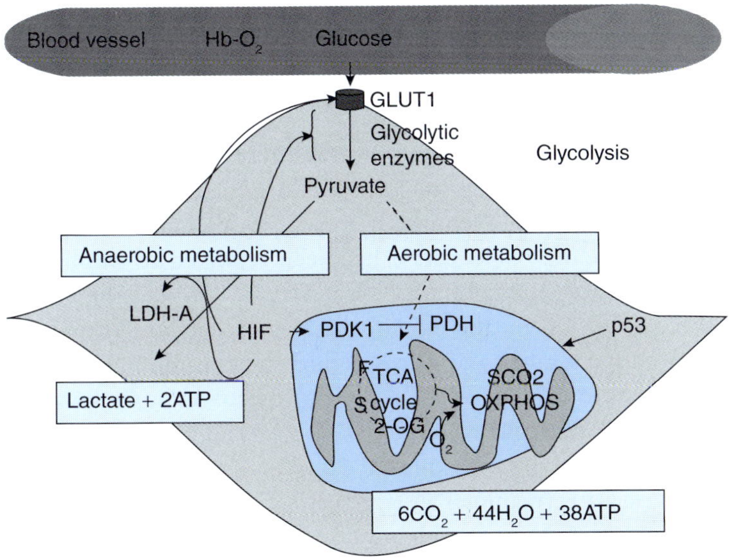

Figure 5. Cancer cell metabolism
Cancer cells opt for glucose metabolism to lactic acid rather than aerobic glucose metabolism to carbon dioxide and water for ATP production, despite the fact that the latter is more energy efficient; producing 19-fold more ATP molecules per molecule of glucose. Hypoxic conditions induce HIF-dependent up-regulation of glucose transporters and glycolytic enzymes involved in conversion of glucose into pyruvate. HIF also tips the balance towards lactic acid production by up-regulating LDH-A, and away from conversion of pyruvate through the mitochondrial TCA cycle and oxidative phosphorylation by up-regulation of PDK1 that inhibits PDH. The TCA components 2-OG, succinate (S) and fumarate (F) respectively, promote or inhibit prolyl hydroxylation of HIF-1α. Certain tumour suppressors such as p53 can tip the balance toward mitochondrial metabolism via activation of synthesis of SCO2. OXPHOS, oxidative phosphorylation.

which limits the production of ATP required for survival and proliferation, tumour cells respond by increasing their capacity to take up and metabolize glucose rapidly. HIF directly up-regulates the expression of glucose transporters (GLUT1 and GLUT3) and of glycolytic enzymes (Figure 5) [5], therefore allowing for more efficient glucose capture and rapid glucose conversion into pyruvate. In addition, cancer cells bypass the oxidative steps for subsequent pyruvate metabolism by converting it into lactic acid [19]. Conversion into lactic acid results in the rapid regeneration of the NAD^+ that is mandatory for glycolysis. Thus, by increasing glucose uptake and the rate of metabolism to pyruvate coupled to diversion to lactic acid, cancer cells can maintain ATP levels under low glucose and low PO_2 conditions. This predilection of cancer cells for conversion of pyruvate into lactic acid contrasts with normal cells that, under normoxic conditions, first metabolize glucose to pyruvate in the cytoplasm and then metabolize pyruvate through the TCA (tricarboxylic acid) cycle and oxidative phosphorylation in mitochondria. In contrast with the truncated cancer pathway, this pathway is oxygen-dependent and 19-fold more efficient in ATP production. However, paradoxically cancer cells, even in the presence of oxygen, prefer to use the truncated pyruvate to lactate pathway, a phenomenon termed the Warburg effect. The mechanism to explain this is not totally clear-cut but probably lies in modification in the activity of oncogenes and tumour suppressors that directly or indirectly affect HIF and HIF-target genes. This might occur through a diverse number of stimuli that implicate signalling pathways including growth factors, nitrogen oxide and ROS (reactive oxygen species), which are known to induce the translation and/or stabilization/activation of HIF, even under non-hypoxic conditions. In particular, the TCA intermediate, oxaloacetate has been reported to inhibit PHDs by binding to the 2-OG (2-oxoglutarate)-binding site [20] and the PHD activity might also be regulated by the oxidative status of the iron bound [21]. ROS produced by mitochondria and released into the cytoplasm of cells exposed to hypoxia promote oxidation of Fe^{2+} to Fe^{3+}. Thus hypoxia would decrease the availability of Fe^{2+} required for PHD activity and might be an explanation for ROS-mediated stabilization of HIF-1α [22]. The PI3K (phosphoinositide 3-kinase)/Akt/mTOR (mammalian target of rapomycin) pathway plays a key role in nutrient utilization through a dual action on glucose metabolism. The Akt protein kinase, when fully active in cells deficient for the tumour suppressor *PTEN (phosphatase and tensin homologue)*, can stimulate glycolysis via enhanced glucose transport and hexokinase mobilization [23], however mTOR can also stimulate respiration [24]. HIF not only directly shunts pyruvate metabolism toward lactate but also inhibits pyruvate metabolism through the mitochondrial pathway. The former occurs through the induction by HIF of LDH-A (lactate dehydrogenase A) expression, the enzyme that converts pyruvate into lactic acid, while the latter occurs though the induction of PDK1 (pyruvate dehydrogenase kinase 1) that inhibits PDH (pyruvate dehydrogenase) [25,26], the enzyme that converts pyruvate into acetyl-CoA. The formation of lactic acid is important for tumorigenesis, as illustrated by the silencing of LDH-A in *neu*-initiated tumour cells that switches pyruvate metabolism from the cytoplasmic to the mitochondrial pathway, with the consequence of reduced tumour growth [27]. These observations suggest that it is possible to harness cancer glucose metabolism in a therapeutic perspective. This is already the case in diagnosis and management where the high rate of uptake of glucose is visualized in tumours by PET (positron emission tomography) detection after injection of a non-metabolizable radio-analogue of glucose [28].

Metabolic intermediates can also modulate the stability and activity of HIF through the regulation of the 2-OG-requiring PHD and FIH dioxygenase activity. The level of 2-OG, or α-ketoglutarate, an intermediate in the TCA cycle, may be limiting in hypoxic conditions (where the TCA cycle is slowed); though it can be produced from metabolism of amino acids. In addition, accumulation of the TCA cycle intermediates succinate or fumarate, which occur in the case of mutations leading to loss of function of SDH (succinate dehydrogenase) and FH (fumarate hydratase) respectively, can act in feedback inhibition [29,30]. In fact SDH, also a component of complex II of the electron transport chain involved in oxidative phosphorylation, and FH like VHL are tumour-suppressors. This information provides yet a further link between HIF and tumorigenesis.

The singularities of cancer cell metabolism also arise from the action of other tumour-suppressors, such as p53, that not only slow glycolysis but also tip the balance of glucose metabolism toward normal mitochondrial respiration. Regulation of glycolysis by p53 can occur though the induction by p53 of the *TIGAR* (*TP53-induced glycolysis and apoptosis regulator*) gene that lowers the level of fructose-2,6-bisphosphate leading to inhibition of glycolysis [31] and through the negative regulation of the glycolytic enzyme PGM (phosphoglycerate mutase) [32]. Regulation of respiration by p53 occurs via the induction of synthesis of SCO2 (cytochrome *c* oxidase 2) [33], a component required in the formation of the mitochondrial cytochrome *c* oxidase complex that is involved in the electron transport chain. Oncogenes promoting cell survival such as Ras [34], Myc [35] and the protein kinase Akt [23,36], are also responsible for the cancer cell predilection for glucose metabolism to lactic acid, in part, through affects on HIF and HIF-target genes. HIF-1 and nutrient deprivation also negatively regulate ATP-consuming protein translation by inhibition of the mTOR pathway [7,37], thus promoting cell survival through energy conservation. On the other hand the activation of mTOR drives glucose metabolism through the mitochondrial pathway and stimulates mitochondrial oxygen consumption and oxidative capacity [24]. It is interesting to note that mTOR is inhibited by the upstream regulator and tumour suppressor TSC1–TSC2 (tuberous sclerosis complex), a finding that provides yet another link between metabolism and tumour suppressor function.

HIF controls tumour pH

The pHe (extracellular pH) of tumours is generally lower than that of normal tissues (6.2–6.8 compared with 7.2–7.4) [38,39]. This is a consequence of the overload in lactic acid and its poor vascular removal but also in production of carbonic acid. Under hypoxic conditions lactate secretion from cells is accelerated through the up-regulation of certain isoforms of the H^+/lactate MCT (monocarboxylate transporter), in particular MCT4 [40], which is located on the plasma membrane (Figure 6). In addition, the carbon dioxide produced diffuses out of cells across the plasma membrane and is converted into carbonic acid by the membrane-bound, HIF-induced ectoenzyme CA (carbonic anhydrase) IX or XII. Thus hypoxic induction of CA IX and CA XII has been suggested to contribute to acidification of the extracellular space (low pHe) [41]. Subsequent uptake of HCO_3^-, a weak base, by members of the Na^+-dependent and -independent HCO_3^- transporters allows for an increase in the pHi (intracellular pH). It is interesting to note that the AE2 (anion exchanger isoform 2) was more highly expressed in VHL rescued, VHL-deficient RCCs compared with VHL-deficient cells, however overall Cl^-/HCO_3^- exchange activity was

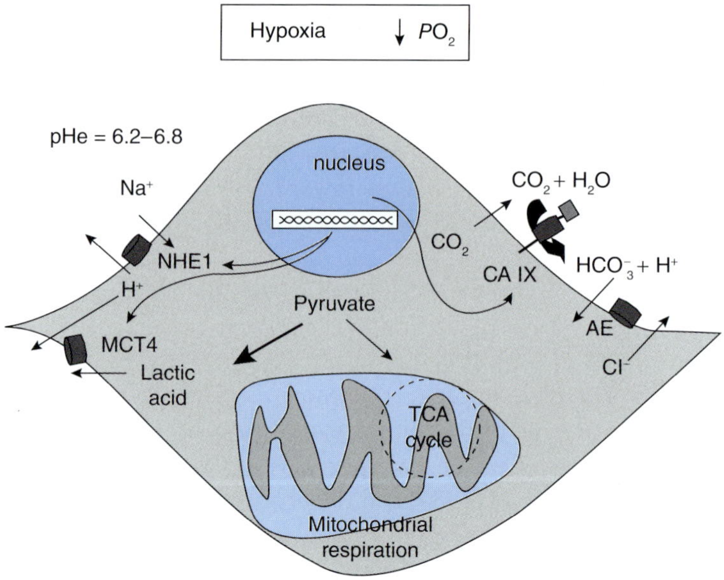

Figure 6. The acidotic nature of tumour cells
The pHe of tumours is lower than in normal tissues due to enhanced production of lactic acid and limited diffusion of carbon dioxide. To maintain cell proliferation HIF modulates the pHi by inducing the expression and activity of the growth factor-activatable and amiloride-sensitive NHE1 and the H^+/lactate MCT4 that are involved in H^+ and lactate extrusion. HIF also up-regulates the membrane-bound ectoenzyme CA IX or XII that converts carbon dioxide into carbonic acid. HCO_3^-, a weak base, is taken up by members of the Na^+-dependent and -independent HCO_3^- transporters, including the anion exchangers (AE), to regulate the pHi.

reduced [42]. Whether these findings point to HIF repression of AE2 expression in RCC remains to be determined. Hypoxia also brings into play one of the most important membrane bound regulators of pHi, the growth factor-activatable and amiloride-sensitive NHE1 (Na^+/H^+ exchanger) [43], the expression and activity of which is increased during hypoxia [44]. CA IX and XII isoform expression is highly induced by HIF and histochemical studies on tumour sections have shown that these isoforms are also highly expressed to the extent that detection has been proposed as a diagnostic marker. In addition to interacting with a number of membrane exchangers and cotransporters the CA II isoform has been shown to interact with MCT1 and to enhance the rate of H^+ flux mediated by MCT1 via a mechanism that does not involve its enzyme activity [45]. It would be of interest to know whether interaction leading to inter-regulation of activity also exists between CA IX and other transporters and exchangers in cells exposed to hypoxia.

The acidic nature of the micro-environment of tumours has also been suggested to promote metastasis through the modification of the activity of HIF-induced metalloproteases responsible for extracellular matrix turnover [46] and through enhanced cell migration [47,48]. Of particular interest is the finding that the HIF-induced glycolytic enzyme PGI (phosphoglucose isomerase) also termed AMF (autocrine motility factor) possesses cytokine activity [49]. AMF/PGI is up-regulated in a number of cancers and the cytokine function has been implicated in cell migration and metastasis. HIF also activates the proto-oncogene *c-met* and the chemokine CXCR4 that are implicated in metastasis [50,51] and represses the epithelial adhesion, tumour suppressor molecule E-cadherin presumably via activation of the transcriptional repressors Snail and SIP1 thereby further promoting the invasive potential of cells [7,52].

HIF in cell survival compared with cell death

Is HIF a pro-survival or a pro-death factor? The answer is certainly both, but with an inclination toward survival, dictated by the stress conditions within the tumour micro-environment at a given time in development of the tumour mass. Specific PO_2 levels, nutrient availability and the pH in the tumour environment might allow cells to adapt and survive through HIF-mediated modulation of angiogenesis, metabolism and pH. Autophagy, a process involving vacuolar trapping of cellular components (ribosomes, mitochondria, protein aggregates) for lysosomal degradation and recycling, might also represent a hypoxia-induced cell survival strategy. However, under more severe stress conditions a cell death response may prevail, though the type of cell death, be it apoptotic or necrotic, needs to be further defined. HIF does induce the expression of several proteins reported to be pro-apoptotic [8]. These include: mitochondrial HGTD-P, Noxa, and the protein BNIP3 (Bcl-2/adenovirus EIB 19 kD-interacting protein 3) and its homologue NIP3L (Nip-3 like) or NIX (protein X). The structural similarity of BNIP3 to other BH3-only pro-apoptotic proteins suggests that it is implicated in cell death. However, BNIP3 seems to require specific nutrient and pH conditions for it to reveal cell death and its precise role in the hypoxic response of tumour cells needs further investigation [53]. Overall, the mechanisms of HIF-mediated apoptosis and implication in events such as autophagic nutritional rescue of cells awaits further clarification.

Conclusions

The recognition that HIF plays a fundamental role in tumour development has stimulated detailed investigation into the mechanisms through which it functions and ways to inhibit its activity as a potential therapeutic approach for cancer [54]. The question as to whether it is best to abrogate HIF activity directly or rather to target specific HIF genes is open for discussion. Indeed, the targeting of the product of the HIF-inducible gene *vegf*, which results in diminished angiogenesis, has shown substantial promise as a therapeutic approach. Exploiting the metabolic predilection of cancer cells may represent an alternative approach in forcing cells to go down the pathway of cell death. However, a better understanding of the role of HIF in cell fate is essential to the development of such an approach.

Summary

- *Hypoxia is a common feature of solid tumours and is associated with resistance to radio- and chemo-therapy, and thus to poor patient prognosis.*
- *HIF is induced under hypoxic conditions and transcriptionally regulates a myriad of genes involved in diverse cellular functions, in particular in cell survival via switching metabolism to glycolysis.*
- *A better understanding of the HIF-dependent metabolic changes that result in extracellular acidosis in cancer cells should lead to the development of additional diagnostic tools and therapeutic approaches that will prove beneficial to the management of cancer.*

The laboratory is funded by grants from the Ligue Nationale Contre le Cancer (Equipe labellisée), the Centre A. Lacassagne, the CNRS (Centre National de la Recherche Scientifique), the Ministère de l'Education, de la Recherche et de la Technologie, and INSERM (Institut National de la Santé et de la Recherche Médicale). We apologize to the many research groups whose work was cited indirectly by reference to review articles.

References

1. Erecinska, M. & Silver, I.A. (2001) Tissue oxygen tension and brain sensitivity to hypoxia. *Respir. Physiol.* **128**, 263–276
2. Vaupel, P. (2004) Tumor microenvironmental physiology and its implications for radiation oncology. *Semin. Radiat. Oncol.* **14**, 198–206
3. Manalo, D.J., Rowan, A., Lavoie, T., Natarajan, L., Kelly, B.D., Ye, S.Q., Garcia, J.G. & Semenza, G.L. (2005) Transcriptional regulation of vascular endothelial cell responses to hypoxia by HIF-1. *Blood* **105**, 659–669
4. Schofield, C.J. & Ratcliffe, P.J. (2004) Oxygen sensing by HIF hydroxylases. *Nat. Rev. Mol. Cell Biol.* **5**, 343–354
5. Semenza, G.L. (2003) Targeting HIF-1 for cancer therapy. *Nat. Rev. Cancer* **3**, 721–732
6. Ferrara, N. & Kerbel, R.S. (2005) Angiogenesis as a therapeutic target. *Nature* **438**, 967–974
7. Pouysségur, J., Dayan, F. & Mazure, N.M. (2006) Hypoxia signalling in cancer and approaches to enforce tumour regression. *Nature* **441**, 437–443
8. Brahimi-Horn, M.C. & Pouysségur, J. (2005) The hypoxia-inducible factor and tumor progression along the angiogenic pathway. *Int. Rev. Cytol.* **242**, 157–213
9. Maynard, M.A., Qi, H., Chung, J., Lee, E.H., Kondo, Y., Hara, S., Conaway, R.C., Conaway, J.W. & Ohh, M. (2003) Multiple splice variants of the human HIF-3α locus are targets of the von Hippel-Lindau E3 ubiquitin ligase complex. *J. Biol. Chem.* **278**, 11032–11040
10. Makino, Y., Cao, R., Svensson, K., Bertilsson, G., Asman, M., Tanaka, H., Cao, Y., Berkenstam, A. & Poellinger, L. (2001) Inhibitory PAS domain protein is a negative regulator of hypoxia-inducible gene expression. *Nature* **414**, 550–554
11. Jain, S., Maltepe, E., Lu, M.M., Simon, C. & Bradfield, C.A. (1998) Expression of ARNT, ARNT2, HIF1α, HIF2α and Ah receptor mRNAs in the developing mouse. *Mech. Dev.* **73**, 117–123
12. Bilton, R., Trottier, E., Pouysségur, J. & Brahimi-Horn, M.C. (2006) ARDent about acetylation and deacetylation in hypoxia signalling. *Trends Cell Biol.* **16**, 616–621

13. Brahimi-Horn, C., Mazure, N. & Pouysségur, J. (2005) Signalling via the hypoxia-inducible factor-1α requires multiple posttranslational modifications. *Cell. Signalling* **17**, 1–9
14. Schofield, C.J. & Ratcliffe, P.J. (2005) Signalling hypoxia by HIF hydroxylases. *Biochem. Biophys. Res. Commun.* **338**, 617–626
15. Kaelin, Jr, W.G. (2003) The von Hippel-Lindau gene, kidney cancer, and oxygen sensing. *J. Am. Soc. Nephrol.* **14**, 2703–2711
16. Peet, D. & Linke, S. (2006) Regulation of HIF: asparaginyl hydroxylation. *Novartis Found. Symp.* **272**, 37–49; discussion 49–53, 131–140
17. Dayan, F., Roux, D., Brahimi-Horn, M.C., Pouysségur, J. & Mazure, N.M. (2006) The oxygen sensor factor-inhibiting hypoxia-inducible factor-1 controls expression of distinct genes through the bifunctional transcriptional character of hypoxia-inducible factor-1α. *Cancer Res.* **66**, 3688–3698
18. Gruber, M. & Simon, M.C. (2006) Hypoxia-inducible factors, hypoxia, and tumor angiogenesis. *Curr. Opin. Hematol.* **13**, 169–174
19. Gatenby, R.A. & Gillies, R.J. (2004) Why do cancers have high aerobic glycolysis? *Nat. Rev. Cancer* **4**, 891–899
20. Lu, H., Dalgard, C.L., Mohyeldin, A., McFate, T., Tait, A.S. & Verma, A. (2005) Reversible inactivation of HIF-1 prolyl hydroxylases allows cell metabolism to control basal HIF-1. *J. Biol. Chem.* **280**, 41928–41939
21. Kaelin, Jr, W.G. (2005) ROS: really involved in oxygen sensing. *Cell Metab.* **1**, 357–358
22. Pouysségur, J. & Mechta-Grigoriou, F. (2006) Redox regulation of the hypoxia-inducible factor. *Biol. Chem.* **387**, 1337–1346
23. Plas, D.R. & Thompson, C.B. (2005) Akt-dependent transformation: there is more to growth than just surviving. *Oncogene* **24**, 7435–7442
24. Schieke, S.M., Phillips, D., McCoy, Jr, J.P., Aponte, A.M., Shen, R.F., Balaban, R.S. & Finkel, T. (2006) The mammalian target of rapamycin (mTOR) pathway regulates mitochondrial oxygen consumption and oxidative capacity. *J. Biol. Chem* **281**, 27643–27652
25. Kim, J.W., Tchernyshyov, I., Semenza, G.L. & Dang, C.V. (2006) HIF-1-mediated expression of pyruvate dehydrogenase kinase: a metabolic switch required for cellular adaptation to hypoxia. *Cell Metab.* **3**, 177–185
26. Papandreou, I., Cairns, R.A., Fontana, L., Lim, A.L. & Denko, N.C. (2006) HIF-1 mediates adaptation to hypoxia by actively downregulating mitochondrial oxygen consumption. *Cell Metab.* **3**, 187–197
27. Fantin, V.R., St-Pierre, J. & Leder, P. (2006) Attenuation of LDH-A expression uncovers a link between glycolysis, mitochondrial physiology, and tumor maintenance. *Cancer Cell* **9**, 425–434
28. Rajendran, J.G., Mankoff, D.A., O'Sullivan, F., Peterson, L.M., Schwartz, D.L., Conrad, E.U., Spence, A.M., Muzi, M., Farwell, D.G. & Krohn, K.A. (2004) Hypoxia and glucose metabolism in malignant tumors: evaluation by [^{18}F]fluoromisonidazole and [^{18}F]fluorodeoxyglucose positron emission tomography imaging. *Clin. Cancer Res.* **10**, 2245–2252
29. King, A., Selak, M.A. & Gottlieb, E. (2006) Succinate dehydrogenase and fumarate hydratase: linking mitochondrial dysfunction and cancer. *Oncogene* **25**, 4675–4682
30. Pollard, P.J., Briere, J.J., Alam, N.A., Barwell, J., Barclay, E., Wortham, N.C., Hunt, T., Mitchell, M., Olpin, S., Moat, S.J. et al. (2005) Accumulation of Krebs cycle intermediates and over-expression of HIF1α in tumours which result from germline FH and SDH mutations. *Hum. Mol. Genet.* **14**, 2231–2239
31. Bensaad, K., Tsuruta, A., Selak, M.A., Vidal, M.N., Nakano, K., Bartrons, R., Gottlieb, E. & Vousden, K.H. (2006) TIGAR, a p53-inducible regulator of glycolysis and apoptosis. *Cell* **126**, 107–120
32. Kondoh, H., Lleonart, M.E., Gil, J., Wang, J., Degan, P., Peters, G., Martinez, D., Carnero, A. & Beach, D. (2005) Glycolytic enzymes can modulate cellular life span. *Cancer Res.* **65**, 177–185
33. Matoba, S., Kang, J.G., Patino, W.D., Wragg, A., Boehm, M., Gavrilova, O., Hurley, P.J., Bunz, F. & Hwang, P.M. (2006) p53 regulates mitochondrial respiration. *Science* **312**, 1650–1653

34. Blum, R., Jacob-Hirsch, J., Amariglio, N., Rechavi, G. & Kloog, Y. (2005) Ras inhibition in glioblastoma down-regulates hypoxia-inducible factor-1α, causing glycolysis shutdown and cell death. *Cancer Res.* **65**, 999–1006
35. Kim, J.W., Zeller, K.I., Wang, Y., Jegga, A.G., Aronow, B.J., O'Donnell, K.A. & Dang, C.V. (2004) Evaluation of myc E-box phylogenetic footprints in glycolytic genes by chromatin immunoprecipitation assays. *Mol. Cell Biol.* **24**, 5923–5936
36. Levine, A.J., Feng, Z., Mak, T.W., You, H. & Jin, S. (2006) Coordination and communication between the p53 and IGF-1–Akt–TOR signal transduction pathways. *Genes Dev.* **20**, 267–275
37. Shaw, R.J. & Cantley, L.C. (2006) Ras, PI(3)K and mTOR signalling controls tumour cell growth. *Nature* **441**, 424–430
38. Cardone, R.A., Casavola, V. & Reshkin, S.J. (2005) The role of disturbed pH dynamics and the Na^+/H^+ exchanger in metastasis. *Nat. Rev. Cancer* **5**, 786–795
39. Gerweck, L.E., Vijayappa, S. & Kozin, S. (2006) Tumor pH controls the *in vivo* efficacy of weak acid and base chemotherapeutics. *Mol. Cancer Ther.* **5**, 1275–1279
40. Ullah, M.S., Davies, A.J. & Halestrap, A.P. (2006) The plasma membrane lactate transporter MCT4, but not MCT1, is up-regulated by hypoxia through a HIF-1α dependent mechanism. *J. Biol. Chem.* **281**, 9030–9037
41. Svastova, E., Hulikova, A., Rafajova, M., Zat'ovicova, M., Gibadulinova, A., Casini, A., Cecchi, A., Scozzafava, A., Supuran, C.T., Pastorek, J. & Pastorekova, S. (2004) Hypoxia activates the capacity of tumor-associated carbonic anhydrase IX to acidify extracellular pH. *FEBS Lett.* **577**, 439–445
42. Karumanchi, S.A., Jiang, L., Knebelmann, B., Stuart-Tilley, A.K., Alper, S.L. & Sukhatme, V.P. (2001) VHL tumor suppressor regulates Cl^-/HCO_3^- exchange and Na^+/H^+ exchange activities in renal carcinoma cells. *Physiol. Genomics* **5**, 119–128
43. Bobulescu, I.A., Di Sole, F. & Moe, O.W. (2005) Na^+/H^+ exchangers: physiology and link to hypertension and organ ischemia. *Curr. Opin. Nephrol. Hypertens.* **14**, 485–494
44. Shimoda, L.A., Fallon, M., Pisarcik, S., Wang, J. & Semenza, G.L. (2006) HIF-1 regulates hypoxic induction of NHE1 expression and alkalinization of intracellular pH in pulmonary arterial myocytes. *Am. J. Physiol. Lung Cell. Mol. Physiol.* **291,** L941–L949
45. Becker, H.M., Hirnet, D., Fecher-Trost, C., Sultemeyer, D. & Deitmer, J.W. (2005) Transport activity of MCT1 expressed in *Xenopus* oocytes is increased by interaction with carbonic anhydrase. *J. Biol. Chem* **280**, 39882–39889
46. Petrella, B.L., Lohi, J. & Brinckerhoff, C.E. (2005) Identification of membrane type-1 matrix metalloproteinase as a target of hypoxia-inducible factor-2α in von Hippel-Lindau renal cell carcinoma. *Oncogene* **24**, 1043–1052
47. Gatenby, R.A., Gawlinski, E.T., Gmitro, A.F., Kaylor, B. & Gillies, R.J. (2006) Acid-mediated tumor invasion: a multidisciplinary study. *Cancer Res.* **66**, 5216–5223
48. Walenta, S. & Mueller-Klieser, W.F. (2004) Lactate: mirror and motor of tumor malignancy. *Semin. Radiat. Oncol.* **14**, 267–274
49. Tsutsumi, S., Yanagawa, T., Shimura, T., Kuwano, H. & Raz, A. (2004) Autocrine motility factor signaling enhances pancreatic cancer metastasis. *Clin. Cancer Res.* **10**, 7775–7784
50. Pennacchietti, S., Michieli, P., Galluzzo, M., Mazzone, M., Giordano, S. & Comoglio, P.M. (2003) Hypoxia promotes invasive growth by transcriptional activation of the met protooncogene. *Cancer Cell* **3**, 347–361
51. Staller, P., Sulitkova, J., Lisztwan, J., Moch, H., Oakeley, E.J. & Krek, W. (2003) Chemokine receptor CXCR4 downregulated by von Hippel-Lindau tumour suppressor pVHL. *Nature* **425**, 307–311
52. Evans, A.J., Russell, R.C., Roche, O., Burry, T.N., Fish, J.E., Chow, V.W., Kim, W.Y., Saravanan, A., Maynard, M.A., Gervais, M.L. et al. (2006) VHL promotes E2 box-dependent E-cadherin transcription by HIF-mediated regulation of SIP1 and Snail. *Mol. Cell Biol.* **27**, 157–169
53. Papandreou, I., Krishna, C., Kaper, F., Cai, D., Giaccia, A.J. & Denko, N.C. (2005) Anoxia is necessary for tumor cell toxicity caused by a low-oxygen environment. *Cancer Res.* **65**, 3171–3178
54. Melillo, G. (2006) Inhibiting hypoxia-inducible factor 1 for cancer therapy. *Mol. Cancer Res.* **4**, 601–605

Subject index

A

B

C

H

I

J

K

L

M

N

O

P

R

S

T